What people are s

Frontlin__

In his wide-ranging journalism and writing, Nick Meynen has been vividly mapping struggles for justice around the world. His new book is a rich collection of the human stories of those struggles – from resistance to mining in India and Greece, to land grabbing in Uganda, to a landmark climate lawsuit in the Netherlands. The book harnesses the power of lived experience to bring our most urgent, high-stakes policy debates to life, and it deserves a wide international audience.

Naomi Klein, bestselling author of *The Shock Doctrine* and *This Changes Everything: Capitalism vs. The Climate*

Prosperity is about our ability to live well on a finite planet. It's a task that has as much to do with story and narrative as it does with numbers and policies. Nick recognizes this essential truth. With the colorful insight of the poet and the dogged persistence of the investigative journalist, he recounts the day-to-day struggles and the extraordinary courage of ordinary people around the world as they engage in the fight for social and environmental justice.

Tim Jackson, author of *Prosperity without Growth*

In Frontlines, Nick Meynen does what we ecological economists and political ecologists often fail to do: communicate our findings in a rigorous but simple language. By reading this book you will learn the same as when reading our academic articles, but instead of falling asleep, you will have fun!

Federico Demaria, environmental scientist, co-editor of *Degrowth. A vocabulary for a new era* and *Pluriverse. A Post-Development Dictionary*

Nick Meynen shares the extraordinary knowledge he has acquired in the last 10 years on environmental conflicts around the globe, from the commodity extraction to waste disposal frontiers. In a lively style he narrates episodes such as the unpaid environmental liabilities of oil companies in the Amazon and the heroic defense of nature and human livelihoods against the sand mafia in India.

Joan Martinez Alier, Emeritus Professor of economics and economic history at ICTA-UAB, 2017 winner of the Leontief Prize and author of *The Environmentalism of the Poor* and *Ecological Economics from the Ground Up*

Nick Meynen's eye-opening journey through some of the worst ecological and social disasters of the world is complemented with something even more urgent today, a chronicle of the inspiring struggles for justice and sustainability that 'ordinary' people are engaging in. This combination of shock and hope is just what is needed to wake up those who are still sleeping through the crises, and inform and inspire those who are awake and responding but may not have the width of vision that such a global tour can provide.

Ashish Kothari, founder of Kalpavriksh, board chairman of Greenpeace India and author of *Churning the Earth: The Making of Global India*

I am sure that this book will be invaluable in helping to create awareness about the destruction being caused globally to provide the materials which make up our modern existence. Most people have no idea where these resources come from and what intense battles are fought to save livelihoods and natural resources in sensitive areas.

Sumaira Abdulali, leading lady of the struggle against sand mafias in India

One often reads in media how we're greening our economy pretty well. But who joins Nick in looking beyond consumerism knows better. He shows systematically how we're plundering the earth's resources – including the livelihoods of communities. If you're still not convinced of the urgent need for system change you need to read this book.

Dirk Holemans, author of *Freedom & Security in a Complex World*

Frontlines

Stories of Global
Environmental Justice

Frontlines

Stories of Global
Environmental Justice

Nick Meynen

Winchester, UK
Washington, USA

JOHN HUNT PUBLISHING

First published by Zero Books, 2019
Zero Books is an imprint of John Hunt Publishing Ltd., No. 3 East St., Alresford,
Hampshire SO24 9EE, UK
office@jhpbooks.com
www.johnhuntpublishing.com
www.zero-books.net

For distributor details and how to order please visit the 'Ordering' section on our website.

Text copyright: Nick Meynen 2018

ISBN: 978 1 78904 192 7
978 1 78904 193 4 (ebook)
Library of Congress Control Number: 2019930340

A CIP catalogue record for this book is available from the British Library.

Design: Stuart Davies

UK: Printed and bound by CPI Group (UK) Ltd, Croydon, CR0 4YY
US: Printed and bound by Thomson-Shore, 7300 West Joy Road, Dexter, MI 48130

We operate a distinctive and ethical publishing philosophy in
all areas of our business, from our global network of authors to
production and worldwide distribution.

Contents

Also by the author

Frontlijnen. Een reis langs de achterkant van de wereldeconomie
(2017)
ISBN 9789462670914

Wandelen met Flora (2013)
ISBN 9789491297519

Nepal. Nieuwe wegen in de Himalaya (2010, 2012, 2016)
ISBN 9789462670693

For Fany, Flora and Rosalie

Mining the St Paul's Cathedral

Two Indians visiting St Paul's Cathedral in London admire the stonework. Ripe for destruction, they conclude. A photoshopped ad in The Times shows a wrecking ball smashing St Paul's up. That day, their friends at UK's Action Aid filed a request to mine the holy cathedral.

This provocative prick to the British establishment is less outlandish than it sounds. The Indians came from their home province of Odisha, India to draw attention to a strikingly similar scenario unfolding there. A London-based mining company, Vedanta, wanted to lay waste to the Niyamgiri Hills to extract bauxite, the raw material for making aluminum. Yet these hills and their primeval forest cover are a sacred spot, home to a pantheon of local gods as well as vital water resources, food, medicine and fuel for the local population.[1]

The St Paul's stunt was part of a 10-year struggle in which thousands of locals had unwillingly found themselves fighting on an increasingly common kind of frontline. Without a fight, forces foreign to them would have eliminated not just their holiest sanctuary, but the key resource in their existence. But they, like many other inspiring fighters celebrated in this book, stood up and gave it a bloody nose. Their struggle could have been the inspiration for the blockbuster movie Avatar, although there are literally thousands of such stories.

The Niyamgiri Hills story is a good place to start, given its richness in angles to view it. Financial giant Société Générale castigated Vedanta's "aggressive plans and misplaced self-confidence" and slapped it down with an eye-watering $7 billion write down in value.[2] The St Paul's salvo even galvanized that bastion of the English establishment, the Church of England,

1

into action. The church delivered a very temporal kind of smite afterwards, by selling up its Vedanta stock. In the end, Vedanta had to give up. Vedanta also lost another frontline battle in May 2018. Their plans to expand a copper smelter in the south of India ended in a closure. But the toll of that frontline was heavy: 22 years of resistance, scores of sick people, 100 days of street mobilization in a row and 13 deaths from police fire.[3] Many of the struggles for environmental justice face repression and violence.

Coming back to the Niyamgiri hills it's interesting to see precisely how little David won from Goliath. A foreign and attacking enemy united Indian indigenous people with the Hindu majority. A coalition of people emerged who normally live miles apart and who do not even speak each other's language. But British anthropologist Felix Padel, who has been following the story for a long time and wrote a book about it, thinks there is a second lesson: "The Indians thought the forest was community-owned and that's why it was impossible for Vedanta to buy pieces of it." The company's commonly used divide and rule strategy didn't work. But a third reason is undeniable as well: cooperation between activists in India and supporters in London worked. The power of multinationals grows, but so does the power of multinational resistance.

* * *

Since 2011, I have been working with hundreds of academics and activists around the world to create an online atlas charting the environmental conflicts that together show that there is a global movement for environmental justice. This Environmental Justice Atlas now contains +-3000 conflicts. We all know even this is just the tip of the iceberg.[4]

What we also know is that these fires have been fueled by an economic policy that climbed to ascendency in the last half century. The conflicts shine a light on a dark underbelly of an economic worldview that applauds the infinite growth of mining, world trade, consumption and gross national product. This muscular beast came of age with the Thatcher-Reagan tandem and by 1989, political philosopher Francis Fukuyama declared that the free market had delivered us to the "end of history". Our souls as well as our material needs were sated, or shortly would be. There would be no more upheaval in humanity's long and rocky road to the good life. Not to the good society, because according to this world view, "there's no such thing as society". Then skyscrapers came crashing down in New York, the financial system went bang and the giddy orators sobered up. Brexit and the election of Donald Trump are surely signs that there are chapters in our history book still to be written, but the resurgence of nationalism and fascism is not the only rewrite that will be needed. I'm convinced that the +-3000 conflicts we have been charting around the world will play a part in that new chapter of humanity's journey on planet earth. The causes of these conflicts remind us that the current political and economic animal is far from stable. To change it, nothing less than a revolution will do. All the signs are there that a revolution based on justice for all, including the natural world, is also brewing.

Visibility of the latter has not been helped by noisy voices who summon alternative facts to call for patriotism, nationalism and white supremacy. The task was already difficult, since the bonds that connect communities across the globe are complex. Our daily routines are irrevocably enmeshed with the ups and downs of rare earth mines in Congo, production lines in Chinese super factories, ebbs and flows within transcontinental gas pipelines and, head in hands, the capricious tweets of one Donald J. Trump.

But we should at least try to separate out the threads of this tangled knot. I will mine down to a seam of understanding, since it is only by understanding the depth of a problem that we can change it. But change also requires telling stories. No revolution ever came about without the stories that moved people into action. Most of our economy starts with mining, so that's where I start my story. Mining is often where the planet butts up against the global economy in spectacular fashion. The second part of the book pulls a pincer movement to look at the other end of the economic pipeline: waste. The third part takes a step back from it all to look at the full picture of the staggering volumes that are passing through the economic pipeline from mine to landfill and to the macroeconomic myths that are busy pushing humanity's journey on earth right towards the abyss. In sum, this book is a storytelling journey along a series of frontlines, a brief analysis of what fuels them and finally some words on what it will take to change course, away from the looming cliff. We sorely need a fairer economy that operates within planetary limits. Few are working harder on this than the well-known and less well-known resistance heroes whose personal histories are at the core of many of the stories in this book. In the fourth and final part, I shine a light on what could and should lead us to a brighter and longer-lasting chapter of the story of humanity on planet earth.

* * *

Who am I to take on this lofty challenge? I am an investigative journalist by training and a subjective journalist through experience. While facts are sacred, journalists are also mere humans and subject to the same process of pre-selection, filtering and coloring of their stories as everyone else. Just as Google's search algorithm pre-selects the results I see according to my previous search history, so my upbringing, education and daily interaction with friends and family chain me in a peskily subjective worldview. Yes, I try to approach every story with an open mind and no story deserves this more than the subject of this book. Carl Bernstein's mantra to "come as close as possible" to the truth will be my bumper sticker. But I equally mark the words of another hotshot journalist, Nick Davies: "Journalists do not have the task of letting each side speak in each story. Look, if source A says to you, the sun shines outside and source B says it's raining outside, what are you going to do as a journalist? Give both versions equal space? That is not journalism. Journalism is to go out yourself and discover who of the two is right." This book wants to do both. I seek truth and I believe that requires going outside, researching and finally choosing the side of those that are closer to the truth than others. I'm not going to waste my and your time on truth deniers. Truth is not the halfway point between a fact and an alternative fact.

This book is not about me so you don't need to know much more about me. Maybe just this: my academic background with masters in both geography and conflict studies as well as my 2 years living and working outside the bubble of Western society prepared me for embarking on the 10-year journey that has led to this book.

Nick Meynen

Chapter 1

Mineral matters

Chasing uranium

In 2011, a professional toxic tour brings me to the legacy of the Bulgarian uranium industry. The ex-mayor of Buhovo walks us – Geiger counter in hand – to the long-closed uranium mine at the edge of the city. At new and nicely sheltered picnic tables, the machine goes through the roof. Stones that fit in a child's hand are a hundred times more radioactive than the normal background radiation. Cows chew on radioactive grass. It's here that people from Sofia often come to on Sundays, eating lunch in between a walk or mountain bike ride. The sun is not guaranteed, but a radiating earth is.[5]

The scientists and activists who follow Todor Dimitrov are united in the European EJOLT project, which investigates and supports the battle of victims of environmental impacts around the world. Also in our international group is the French nuclear engineer Bruno Chareyron. Bruno's job is to perform independent radiation monitoring in order to improve information and protection against radiation. While using his own devices, Bruno tells us that there are still problems in France with the inheritance of more than 200 closed uranium mines. After a long campaign, the big French company AREVA was obliged to clean up dozens of sites. But in Bulgaria, the damage of the nuclear industry isn't even mapped yet. Bruno hopes to help change that.

Our group of uranium disaster tourists follows the route the uranium used to take: from Buhovo's uranium mine, which still contains about 12,000 tonnes of uranium, to the processing plant on the other side of the city. There, uranium-containing rock blocks were converted into yellowcake, which then went to Russia to make nuclear fuel. The liquids needed to separate the uranium from the rock remain radioactive, almost into eternity. They lay in dark puddles and pools, behind a clumsy dam next to the factory. Over 10,000 people live directly downstream. But if the dam breaks, the radioactive sludge would draw a trace from Buhovo across the plains around Sofia all the way to Romania.

To avoid a nightmare scenario, the European Union (EU) paid the Belgian company Bitumar Soils Joint Venture 3 million euros to strengthen the dam. But after a downpour in 2009, local residents claim that a piece of the dam came loose. "We did not get permission to check that up close," says Bulgarian environmentalist Todor Slavov. We too can't get closer than 100 meters from the dam. Slavov points to a patrolling pick-up truck between us and the toxic sludge and tells us that it is dangerous to approach closer.

The inheritance of the mine, the factory and the toxic pools require a quasi-perpetual cash flow to prevent a disaster. Is there going to be enough money in 10 or 100 years? The yellowcake factory has been closed since the turn of the century, but the promised decommissioning has not yet taken place. Guardians block access to the complex, a complex that Dimitrov believes is in the hands of a new private investor. He adds that nobody in Buhovo knows why a private business invests in a heavily polluted rusting former uranium plant, in a country that has had no working uranium mines since 1992. When our minibus pulls up to the gates, the armed guards need no words to make it clear that we're not going to find out now. I quickly shoot a few pictures but the driver smells trouble and makes a U-turn. There are questions here that you can't even ask.There are plenty of questions. Who will and who should pay for the quasi-perpetual costs associated with the management of the radioactive wastes left after uranium production? Taxpayers in the country that bought or the one that dug the uranium? European citizens or a private company?

I wonder if Buhovo's uranium ever reached the nuclear

power stations that still drive my commuting train to Brussels. To my surprise, I learn that the reverse road is more likely: "The nuclear power station in Bulgaria uses uranium that you used first." At least, that's what Georgi Kotev says, a nuclear physicist and former employee of the Kozloduy nuclear power station. He became known as the whistleblower of the Bulgarian nuclear industry. Kotev: "I calculated the time between two refuels of our reactor. When I noticed something abnormal, I was told to adjust the software so that everything would look normal. That was suspicious so I looked into the safety report and it showed that we had changed fuel without the normal procedure being followed. Since I knew too much I was fired and ever since I get threats." If I ask Georgi what the reason would be for secretly changing the kind of fuel. Georgi:

> Money. Bulgaria still pays Russia the new fuel price, but in practice we get the cheaper, more risky and more radioactive recycled fuel. Probably fuel you've used in the West and exported to Russia for processing. Following my complaint with the International Atomic Energy Agency, a so-called independent investigation was instituted under the supervision of a Bulgarian who pre-selected the fuel to be checked. Look, there are powerful people behind this, who now carry a defamation campaign against me.

Bulgaria is still in the EU, so the country has environmental standards to comply with and a watchdog above it. This makes the extraction of uranium more expensive because it has to invest to protect people from poison and death. Therefore, it is cheaper for the nuclear industry to exploit uranium in Namibia and import it. The boss of the Australian uranium producer Paladin, who is active in that country, is honest about that. He said that Australian and Canadian environmental and social standards are over-sophisticated. But even the much tighter

European standards are not protective enough, according to CRIIRAD. One example: the norm for the presence of radioactive tritium in drinking water is a factor 70 stricter in Europe than in Canada.[6] But in Canada there's at least a norm. In countries such as Namibia, there are simply no standards.

If we continue to put non-European uranium in European nuclear power stations, we don't resolve but export uranium extraction problems. Whose deaths are the deaths from pollution around uranium mines in Africa? It's a question few nuclear energy proponents dare to ask.

Bertchen Kochs – the Namibian activist in our group – asks these questions all the time. She says that mining in Namibia consists of open pits from which the wind blows radioactive matter across the national park. France is full of uranium and depends on nuclear energy like no other European country, but it now extracts all its uranium from foreign countries such as Niger. Much mining and also dirty industry moved from Europe in the last half century. This applies not only to uranium, but also to digging coal or producing steel or pesticides. All this was encouraged by instances like the one at the World Bank. Lawrence Summers was the chief economist of the World Bank when he spoke about the need to move dirty extraction and production to the under-polluted countries. He used that term in a leaked internal note. The cited reasons for moving polluting mining and industry were that wages and standards are lower and that the deaths by pollution are less costly.[7] Lawrence Summers, in addition to being director of Harvard University, was also a key adviser to various US presidents. But while he became infamous for that leaked memo, his opinion was and still is the mainstream opinion that steers policies in that direction. Of course, in language for the general public this becomes "efficiency gains" and "bringing development to poor countries".

* * *

Years after our Bulgarian toxic tour I'm still in touch with Bruno Chareyron. His cool scientific approach to a controversial political theme arouses my curiosity. Bruno loathes ideological or emotional debates, he rather works with facts and evidence. In those days we spend on toxic tour in Bulgaria, he makes me doubt about the dividing line between scientists and activists. Maybe Bruno is a *factivist*: an activist for whom only the facts count. The story of the person Bruno is therefore hard to find. And yet I wanted to know who the man is that the newspapers in Africa write about when he arrives with his appliances to measure radioactivity. And especially: how he came to that point. The facts about Bruno himself can be short. Bruno is a French engineer in energetic and nuclear physics with a master in particle physics. For over 25 years now he's been the director of the CRIIRAD Laboratory (CRIIRAD).[8] CRIIRAD was founded as an NGO in 1986 by disappointed French citizens. Their government lied about the actual impact of Chernobyl on their country – just like Belgium's weatherman Armand Pien had to lie when he gave the weather report on the day of the disaster. In order to be able to do independent measurements, CRIIRAD started its own lab. Soon, civil society associations who suspect a serious impact of nuclear industry in their environment found their way to the professional team. Bruno studied radioactive contamination in Europe, Africa, Brazil and Japan. After the disaster in Fukushima, he traveled to the disaster area. I ask him what he thought about nuclear energy when Chernobyl exploded:

> I did not have a clear opinion on nuclear energy at the time, only an intuition that this energy form would be too strong to be controlled by human beings. At the time I was unaware of fraudulent lobbies. After the Chernobyl disaster, the French

government lied. By stating there was absolutely no danger in France. At the engineering school, the professors said that too. I was curious to learn more about it and worked for 2 years as a researcher for EDF (Electricité de France) on the protection against radioactivity and reactor physics. Later, I learned at the French Embassy in Ireland to work with renewable energy sources, after which I started working for a company that made solar cells.

When it comes to energy, Bruno is both a generalist and specialist. His driving motive: learning about the best energy source for all the energy needed by humanity. Bruno's son was 2 when the family found that he had leukemia. They lived and worked in Normandy, not far from the AREVA processing plant in La Hague, on the north French coast. That plant treats radioactive waste and dumps large amounts of radioactive substances into the air and the sea. In the hospital, Bruno discovered that people were saying that children living close to the factory had a higher risk of leukemia. There and then he decided to apply for research into the impact of radioactivity on the environment and health. At CRIIRAD he started a long-term field research that has been ongoing for a quarter of a century. In 2016, Bruno received the Nuclear Free Future Education Award for his extraordinary efforts.[9] I ask what his classmates do and what they think of his work. Most seem to work for the nuclear industry, and according to Bruno, most are pleased that there is something like CRIIRAD. They feel that there is a need for companies and government to keep an eye on safety and pollution, "but most of my colleagues working for the nuclear industry know little to nothing about radioactivity and even less about the whole chain. They also know nothing about the conditions of the environment around the places where the uranium is coming from." One result of globalizing the chain from uranium mine to nuclear energy is that both the end users and professionals at the end of the chain have lost touch with the realities at the start of the chain. When I ask Bruno on what energy his inner engine runs he immediately responds with the lies distributed by the nuclear industry. But he also says he hates these so-called experts, who feel better than Joe Sixpack. Suddenly, we're talking about his Protestant background, where one learns that every person is unique and must be able to form his or her own opinion.

That culture was also nurtured in the college where I went to school. The good thing about our current work is that we help communities to better protect their rights. We provide them with the knowledge and arguments they need when they face companies and government. They need that info to better protect themselves from damage to their environment and health. In doing this work I made connections with people from France, Niger, Mali, Japan and so many other countries.

Bruno's most emotional memory of his fieldwork is a conversation with an old Japanese lady.

We had measured the radioactivity on her farm, near Fukushima. The government had said that they could return, but our measurements showed that this would be very unhealthy. With our device, we were able to show her how wild the Geiger counter went on her field. It is thanks to these measurements that she decided to stay away from her former farm. She told us, "You've come from far to help us. I was in the darkness and you brought the light in." She looked a little like my own grandmother. The feeling I then had cannot really be described.

Bruno changes subject. How frustrating is it to be in the business of bringing bad news? To say to people: you cannot live here anymore. Bruno admitted that he sometimes struggles with it, even after 25 years working for CRIIRAD. "Promoting renewable energy would be a lot more fun than always investigating the negative impact of the nuclear industry. But I know that in many places we have been able to improve people's protection. With our campaigns, we have been able to force companies to clean numerous sites in France, Niger and other countries. We can't stop the whole industry, we cannot do that alone. But helping people to not get sick and giving them the information they are entitled to, that's enough motivation for me."

* * *

A year after my interview with Bruno, I join him on his visit to the European Parliament. His goal: to expose AREVA's myths about the impact of its uranium mining activities in Niger and call for help at the European Commission. AREVA's own study showed that 16 percent of a sample of houses in a city of 200,000 are built with radioactive material. According to Bruno, people living in the area are exposed to radioactivity and death rates from respiratory diseases twice as often as in the rest of the country. Bruno, Almoustapha Alhacen from a local NGO in Arlit, and Michèle Rivasi, a French member of the European Parliament from the Green faction, tried to convince people from the European Commission to at least go on a fact-finding mission in Niger. It all seemed to fall on deaf ears.[10]

France and Belgium are entirely dependent on imports of uranium and derived products. We do not extract uranium, but half of our electricity comes from nuclear power. Through AREVA, for example, we get our stuff from Niger, where a lot of people live without flowing water or access to electricity but with a radioactive environment. Even within the well-regulated EU, there are cases of secret deals in radioactive fuels and horror stories around radioactive waste. The precise cost tags are unpredictable, but the one thing that's sure is that they are almost ad infinitum. Outside the EU, the situation is even worse. Or was it a coincidence when French troops went to Mali in January 2013, just when some rebels came near to AREVA's most important uranium mine in neighboring Niger?

* * *

My home country, Belgium, depends on the French nuclear industry. We never debate about the problems in our supply chain, but we do debate the risks at home. The Netherlands, Luxembourg and Germany have all asked Belgium's government to close Belgium's most risky reactors with immediate effect. Thirty cities are suing Belgium for not closing them. Here's why.

After pushing a long due retirement back from 2015 to 2025 it took 2 days into 2016 before the first incident took place in one of Belgium's nuclear power plants. Back in 2012 it became known that the mantle around the old Tihange 2 reactor showed signs of erosion. Further research in 2015 concluded that there are thousands of cracks of up to 15 cm in the mantle. Later that year, ten security incidents were recorded in Tihange in just 6 weeks, leading Belgium's nuclear safety agency to suspend four members of staff and raise serious questions about the safety culture. In 2015, Belgian's nuclear plants spent longer in shutdown or "maintenance" than in being operational. Despite all that, Belgium's government decided to postpone the already agreed retirement in 2015 by 10 years. But if Tihange is a basket case, it is the Doel plant that really reads like a horror story. There are a staggering number of cracks in the mantle that is supposed to keep the Doel 3 reactor in check: 13,047. The cracks are on average 1 to 2 cm wide, but the largest ones are up to 18 cm. And with 35 years of operational history, the researched Doel 3 is the second "youngest" of Doel's four reactors. Belgium's nuclear safety agency concluded after the tests in Tihange and Doel that the erosion of the mantle was due to normal reactor activity. But the German government no longer trusts the Belgian Nuclear Safety Agency.

Ageing is one worry, terrorism another. Doel was sabotaged in 2014. The culprit(s) remain unknown. In 2015 police found hidden cameras that followed the movements of a nuclear researcher, raising questions about criminals extorting staff. France has already experienced a series of undeclared drone flights over various nuclear power stations. The Bulletin of the Atomic Scientists warned that drones can easily carry AK47s and drop them inside the territory of the plant at night. They also explained that drones can attack the power lines and then the diesel generator back-up system. Belgium also seems to be a great hide-out for terrorists.

In terms of potential impacts, Doel is by far number 1 in Europe. The major Fukushima disaster knocked 2 to 10 percent from Japan's Gross Domestic Product (GDP), but when Doel goes into meltdown, the cost is estimated to be 200 percent of the GDP of Belgium. Not that GDP will matter at that point. Most of Flanders would become an inhabitable zone, sending millions of refugees to France, the Netherlands and Germany. That is, if they don't close the borders.

Policymakers are taking this gamble because they and their voters are not experiencing any of the real-life consequences of nuclear energy yet. To speed that process up, I have a suggestion for Almoustapha Alhacen: next time you visit Belgium, take a few bottles of so-called drinking water from Arlit with you. Go to a popular place in the port city of Antwerp, ideally a place from where the Doel nuclear power plant is visible in the background. Take a TV news crew with you. Address all political leaders in Belgium who decided to postpone a long overdue retirement with another 10 years, thus guaranteeing continued mining of uranium. Just look into the lens and invite them to drink the bottle with you.

Gold fever in Greece[11]

Greece has been reeling under the spread of TBC, malaria and malnutrition. According to The Lancet, the medical journal, Greece is experiencing a humanitarian disaster and, according to the authors, this is a direct consequence of the policies imposed by Greece's creditors.[12] They demanded a criminal halving of the pharmaceutical expenses. The pensions also halved within a few years and for half of the youth there is simply no work on offer.

Greece is grounded, while the ground is full of gold. If you judge on that basis, Greece is a treasure trove. The solution seems simple: get the gold out and make the Greeks rich. The reality is anything but simple. At the end of 2016 I went to some Greek gold mines in construction, to what is a relatively hot frontline for European standards.

This story has a very long history, but I'll fast forward to January 2015. The left party Syriza wins the elections and makes a government in record time. A key promise was to cancel the planned gold mines. Syriza started by telling Hellas Gold, daughter of the Canadian mining company Eldorado Gold, to redo their homework to get the proper licenses.

After a year with Syriza in power, Eldorado Gold wrote off more than 1 billion dollars of expected revenues, after which the company's course crashed. Between the arrival of Syriza at the end of January 2015 and the end of January 2016, Eldorado Gold's share fell from 8 to 2 dollars. But in the spring of 2016 a number of secondary permits were given, through the court. The main license, for the gold and metallurgy factory, is still missing, but Eldorado Gold is still trying to get it.

Enough poison in the ground to kill everyone on earth

Officially, the issue is simple: can Hellas Gold convince the Greek government that flash melting with Halkidiki ores is safe? This method of separating gold from ore (as a by-product in the mining of copper) is common. The procedure is in the environmental permit of the project. The only other option, using cyanide, was once granted to a previously disastrous mining company (TVX, also Canadian) but never used. Why? Because the use of cyanide was later prohibited by the highest Greek court due to the expected environmental impact. So only flash melting is an option.

Arsenic. That summarizes the problems of Hellas Gold. Ore of the planned Skouries mine would be melted together with ore of the nearby Olympia mine, rich in arsenic. The process will produce 20,000 tonnes of arsenic. Annually. The arsenic content in the soil is 16 times higher than that allowed for flash melting in China and is almost 30 times above the EU average. The total amount of arsenic would be sufficient to deliver a deadly dose to every human being on earth. Three times.

How the Greek government can be convinced that all this can be done in a safe manner is unclear – not least because Hellas Gold tried to cheat earlier on, sending the government false ground samples. The anticipated storage for all that arsenic is mind-blowing. A seismological expert of the Aristotelian University completely destroyed the so-called earthquake-resistant character of the 150-meter-high dam behind which the lake of arsenic should come.[13] A tectonic fracture line with a lot of seismic activity lies right below it. In 2002, acid mine waste water, laden with heavy metals from the Stratoni mine, was discharged into the Stratoni bay, coloring it red. Swimming and fishing in the bay have been forbidden since 1980 because of mine pollution.

Despite all this, Hellas Gold is already storing arsenic wastes from other mines on that earthquake prone site, although the dam itself is not finished. Putting the cart before the horse seems part of a tactic. Hellas Gold also used a license for merely placing heavy machines to build the foundations of the metallurgy plant, to cut very large pieces of forest and to top off a mountain. During a visit on October 17, 2016, a flurry of activity by large cranes, bulldozers and lorries made loud noises where previously only some birds were singing and bees were buzzing. The dust made today is peanuts compared to the dust that will be released when the mine is operational. Eldorado Gold itself estimates that 2162 tons of dust will spread around the area. Per hour. That seems impossible and is probably due to a typo error in their report. But even if they meant 2162 tonnes of toxic dust per year, the unit used elsewhere in their report, it still implies that the bees, olive trees and sheep in the wide area will be covered with arsenic poison, putting the whole bio-economy out of business.

Enough natural wealth above the ground to give everyone a good life

The bees in this area are not ordinary bees. They make the best honey in the world, witnessed by a beekeeper from nearby Arnaia who received this honorary title for his product. The 1059 beekeepers in Halkidiki are an export economy in their own right. The whole sector is threatened by the mining plans. Idem for the many olive producers, shepherds and the tourism industry. Thousands of jobs will be lost. Giorgos Karinas (41) is such a beekeeper. He survives in Megali Panagia, the nearest village to the planned Skouries mine. Karinas says he would not agree if I wrote "lives in Megali". He says his life is at a permanent pause. "We live in eternal uncertainty, for so many years. I have bees and make olive oil, but I cannot invest as long as I do not know if the mine comes. If their toxic substance and water ruin this village, how can I still sell my honey and olive oil?" We take a small tour of beautiful places with tourist value and we drink from what is now still healthy spring water. He turns melancholic when talking about the resistance: "In 2006, 5000 people signed the petition against the mine, more than the entire population of this village! But in the 2008 crisis, many lost their jobs in the construction sector, and some began to work for Hellas Gold. Now the village is divided. I myself think of emigrating." That radical option is shared by many people, though nobody knows where to go. It's a sad decision in a rich region, a region with a lot of renewable natural capital like forests, bees and beaches.

And then came Canada

Giorgos Vlachos (50) is the latest shepherd in a long family line – and probably also the last. His village, Palechori, had 20 dozen shepherds who made Greece's famous feta cheese. Giorgos: "In this village, fathers now tell their daughters to marry a miner. This village chooses temporary jobs and ignores asbestos." For a long time, Hellas Gold kept secret that the soil is also full of asbestos – until the geological institute confirmed and published it in January 2016. Everybody here now knows it. When you enter the neighboring village of Megali Panagia, a big banner welcomes you by saying: "Asbestos is a slow killer, but don't forget: he's coming."

There are more things Hellas Gold is trying to keep secret. Vlachos tells about what happened to two colleagues with herds close to the Olympiada mine: "Their milk was so full of heavy metals that they could no longer sell it to the company where I know the boss. Through him I know that they got silence money from Hellas Gold. The company later bought their herd and is trying to convince all shepherds to sell their sheep and goats to them."

Suddenly and to my great surprise, this shepherd in jeans, which were probably not worn out by design but by hard labor, begins to talk to me about a free trade agreement: the Comprehensive Economic and Trade Agreement between Canada and the EU (CETA). At the time, CETA could get the green light at any time. "When they sign CETA it is game over for me." My first thought: The Greek government has issues with

Canada for making feta cheese, a Greek specialty that Canada wants to produce on an industrial scale. But that's not what this sheep shepherd has on his mind. "If CETA comes, the mine will come and if the mine comes, I can close my books." How he makes that connection? Well, the gold mining planned by the Canadian company Eldorado Gold in the region would destroy the conditions for Vlachos's business to thrive. CETA will make it extra hard for any Greek Government to pull out of the mine – because they then risk a fine to the tune of billions, which they then most likely will have to pay.

* * *

Back in the regional capital of Thessaloniki, Maria Kadoglou (48) taps into her 19 years of experience in this struggle to give me more elements of this frontline story.

It's a guess how much compensation they would enforce through CETA, but look at Romania. The Canadian mining company Gabriel Resources demands 4 billion euros of the state for not allowing the Rosia Montana gold mine. The company is using the arbitration mechanism of a bilateral trade agreement between Canada and Romania that is very similar to what is foreseen in CETA.[14] And in Romania, they did not even get to the stage of an approved environmental impact assessment. Hellas Gold claims that they already spend 700 million euros on the Skouries gold mine in construction and that the potential profit is 10 billion euros, so the fine they can ask thanks to CETA will be somewhere in between.

Maria also points to other gaps in my story: "The European Water Directive was also violated here. The former Greek government just gave the whole mining region an "exception" status. WWF Greece submitted a report to the European Commission at the beginning of 2015, but they did not respond."[15] The Kakavos forest was threatened by the gold mine. Kakavos means "the mountain that can never burn". That's because there is so much water in the ground that the forest never dries out. It is the most important freshwater source of the whole region. There is no way the mine can be made without destroying that.

Who benefits from whom?

The biggest creditors of Greece, gathered in the troika, have big plans with the country. The troika consists of the International Monetary Fund (IMF), the European Commission (EC) and the European Central Bank (ECB). The loans they give to Greece also lay down strict conditions in addition to interest rates. The EC intends to improve the entrepreneurial climate. The IMF claims that it wants to attract foreign investment again. Therefore, in 2013 Greece had to sign a memorandum that has a curious passage in it: "To facilitate investment...we will... streamline...licenses and permits (operational, environmental, land use...) by reducing their number."[16] With the troika knife at its throat, Greece signed what in practice boils down to a pledge to circumvent European environmental legislation. For the IMF, this is not entirely new. For many decades, the IMF has imposed similar conditions on developing countries when "helping" them with loans. For decades, the IMF paved the way for the big mining companies to do cheap and harmful mining – free from costly environmental and social protections. They've done that in what they call the "under-polluted" countries. The novelty is that this is now happening in the EU.

It helps the troika that Greece is no longer a democracy. The creditors will not put it that way, but anyone looking at the facts can hardly claim that Greece is a democratic country. Rituals such as elections and referenda are still there to keep up appearances. But during the Greek elections in January and September of 2015, government leaders in Europe scrambled over each other to shout that Greek voters can just as well stay at home. Because, so they said, whatever the result will be, the reforms and conditions imposed on Greece – which define virtually ALL aspects of their policies – will in any case continue. People in the country where the word democracy was invented sure know that democracy is something else than that.

Take the former Dutch Minister of Finance Jeroen Dijsselbloem, the chairman of the Eurogroup, which brings together all finance ministers in the eurozone. He and his German colleague Wolfgang Schäuble said in their first meeting with their new Greek colleague Yanis Varoufakis that he just had to carry out what had already been decided by the elected representatives before him who were just punished by voters in the most crushing way possible. The two most important people of financial Europe added that they will destroy the Greek banks and thus the distribution of money in Greece if he didn't. Varoufakis calls this "fiscal waterboarding". In the summer of 2015, the Greek people rejected the troika's proposals in a referendum. The troika reacted by adding insult to injury plus extra pressure until the Greek government agreed with what the people had just rejected. You can compare that to holding a European summit 10 days after the Brexit referendum, in which the UK's Prime Minister is held hostage in Brussels until an agreement for closer cooperation with and more payments to the EU is signed. Under the EU's conditions. Unimaginable? What happened to Greece was worse.

How the European elite plunders the Greek state treasury

Jeroen Dijsselbloem and Jean-Claude Juncker are both wearing two hats in this. The Netherlands and Luxembourg are the two largest foreign direct investors in Greece. Strange, for such small countries far from Greece. Until you realize that 80 percent of Dutch investment in Greece comes from companies that only have a mailbox in the Netherlands. The Dutch Centre for Research on Multinational Corporations (SOMO) discovered that Eldorado Gold has at least 12 mailbox firms at one address in the Netherlands. Just one of these 12 companies avoided paying nearly 2 million euros to Greece.[17]

It's made complex on purpose, but the route the money follows is pretty simple. This is how it works. Hellas Gold, a subsidiary of the Canadian company Eldorado Gold, finances its activities in Greece through the issuance of bonds. A mailbox company based in the Netherlands buys that. A Barbados-based company buys that off the Dutch mailbox company. And that company in Barbados is owned by Canadian Eldorado Gold. The Canada-Barbados-Netherlands-Greece detour has one goal: bypassing the Greek state treasury. Both the CEOs of Eldorado Gold and Hellas Gold have admitted to avoiding paying taxes, adding that it was a legal practice and that everybody did it.[18]

In the Netherlands, these mailbox companies flourished thanks to Jeroen Dijsselbloem, who was the Minister of Finance from 5 November 2012 to his party's historic election defeat in March 2017. Tax avoidance is the only function of these mailbox companies. The Netherlands has some 20,000 of them that together offer a detour from state coffers for around 4000 billion euros each year. The defunding of states to pay for schools and hospitals and roads is called "optimalization", by the likes of Dijsselbloem, a caviar socialist.[19] The Netherlands has a world record of this kind of tax evasion in its name. That brings me to the key question in all this: with what moral authority does Dijsselbloem tell his Greek counterpart to get its finances in order?

The President of the European Commission, Jean-Claude Juncker, doesn't do any better. As Prime Minister of Luxembourg, he also stimulated the prosperity of mailbox companies in his country and that has also come at the cost of the Greek state. LuxLeaks alone showed that Greece lost billions in Luxembourg thanks to Jean-Claude Juncker. As President of the European Commission, Juncker sells a story of solidarity and sympathy with Greece, that naughty boy from his class that had deviated from the right path, but that he, Jean-Claude, was bringing back on track again. When the Greek Prime Minister Tsipras came to Brussels and held a press conference with Juncker, the latter was giving him little slaps in the face, literally. The body language was obvious: you've listened to daddy and now you're a good brave boy again.

Syriza: mosquito bites on a mining giant

Before Syriza came to power in 2015 the mining business in Greece was on a roll. Violations of building regulations, hazardous waste legislation, heavy metal pollution in rivers: none of that stopped the urge to get at the Greek gold. In 2011, former Minister of Finance George Papaconstantinou became Minister of Environment. He immediately gave Hellas Gold its missing license. Later, he tried to delete names of his relatives from a list of Greeks with Swiss bank accounts. In 2014, the Greek Parliament tweaked forest law during the summer recess. At the eleventh hour a clause was added which allowed the construction of a crucial part of Eldorado Gold's mining infrastructure in a pristine forest. For journalists and nature conservationists, the Greek law is applied with a lot more vigor. Articles and tapped telephone conversations from journalists are used as "evidence of criminal activities". The journalist Kostas Vaxevanis published who evaded taxes in Greece on the basis of information given to the Greek government by Christine Lagarde, the IMF's boss. Not the tax evaders, but the journalist was brought to justice.[20] In total, more than 450 anti-mining activists have been charged, often just for participation in a protest march. For 38 of them, prosecutors are asking for a 30-year jail sentence for "participation in terrorist activities". Sometimes, penalties for the real criminals do materialize. Former Development Minister

Akis Tsoxatzopoulos, who signed the concession for Hellas Gold, was arrested for money laundering and on 7 October 2013 he flew behind bars for 20 years. Former Deputy Minister of Economic Affairs, Christos Pachtas, was taken out of office after he converted 17 hectares of protected primeval forest around Halkidiki to luxury homes. But the real top, those that make the draining of the Greek state treasury possible like Dijsselbloem and Juncker remain untouched. They're not even considered to do harm. Draining the Greek state coffers is considered a bonafide thing to do.

Since Syriza is in power, there's new hope that Halkidiki will receive the protection that the area deserves under Greek and European laws. In January 2016, the Ministry of Environment gave Hellas Gold a fine of 1.7 million euros for 21 environmental offenses, such as a discharge of heavy metal wastes. The permitted arsenic standards were already exceeded at several water sources. Incidents such as these were never reported. But what is 1.7 million euros? If the mine is finished, Hellas Gold expects the first 7 years, 4000 and the next 20 years, 2550 kilos of gold. At the early 2017 gold price, that brings the total turnover to almost 3 billion euros. Another mine, in Olympia, should create a similar revenue from 2020. In total, Eldorado Gold hopes to generate about 10 billion euros in Greece. For the moment, nothing really guarantees that even the anti-gold mine party Syriza will be able to do more than make some mosquito bites on a mining giant. And even that seems difficult to implement. In January 2018 the fine was annulled by an administrative court.

* * *

The Greek professor Giorgos Kallis is a political ecological economist. With his many peer reviewed publications, he is a leading voice in the growing discipline of alternative macroeconomics. Kallis tells me that what Greece is experiencing today is a regression from a developed country to a developing country, similar to the process that many Latin American countries passed through in the 1980s and now continue to be in. "The sole function of a developing country is to provide the global economy with cheap raw materials, often at the cost of its own people and its own development." Kallis says it is not a coincidence that the gold – which has always been there – is dug out in this time of crisis. The crisis lowered costs by reducing the cost of labor (25 percent) and reducing the monetary cost of externalities: the health, visual or environmental impacts are no longer valued thanks to the troika conditions. Kallis: "*Economic crises are necessary* for creating new exploitable territories when limits have otherwise been reached. They achieve it by the devaluation of economic, social and environmental capital. Having a crisis like Greece is having creates an economic opportunity for the gold miners."

Kallis goes one step further than most of his countrymen. He says that the crisis is a prerequisite for profitable gold mining in Greece. And according to Kallis, people are actively working to meet that condition in spaces outside of Greece. He said that before WikiLeaks published in April 2016 the transcript of a phone call in the IMF. In that call, two of the IMF's top officials, Poul Thomsen and Delia Velkouleskou, talk about the best timing for the next Greek credit crisis.[21] You can of course also interpret it as an open-minded conversation between people who see the inevitable: that Greece cannot repay its debts. Even a top figure inside the troika told the Greek Finance Minister Varoufakis, in a private conversation, that the program imposed on Greece "could not work". Indeed, for the Greeks it does not work. But for the Canadian Eldorado Gold and for large German companies – taking over the obligatory privatized regional airports and urban water facilities – it works very well.

Mothers with courage

The anti-gold movement wants the official revocation of Eldorado Gold's activities around Halkidiki, restoration of the environmental damage in the region and the declassification of the region as a mining region. Expressing these demands is not without risk. On March 20, 2012, 30 members of the community who held a peaceful sit-in were beaten up by a mob of Hellas Gold employees, with the assistance and encouragement of the management. Eight locals had to go to hospital. In October 2012, the police attacked and chased peacefully protesting women for a full 7 kilometers – enough for Amnesty International to launch a human rights violation case against the Greek state. In March 2013, more than 200 well-armed police attacked the village of Ierissos, known for its resistance to the mine. Teargas bombs ended up in homes and in the local school, where a student was seriously injured. Dozens of people needed first aid, including a baby. In reaction, 20,000 people marched in Thessaloniki. Towards the Canadian Consulate.

Halkidiki is not the only Greek frontline that suffers from gold fever. Years ago, I came into contact with Elena – a fake name that I use to protect her identity, on her insistence. Elena lives in Kilkis. A mining company wants to transform a large fertile area where 15,000 people live into a poisonous pit. Against the will of the majority. Elena emphasizes that she is not a tree hugger. "When it first learned about the proposed gold mine, I was totally in favor. I was annoyed by the people who had reservations about the plans because I found that finally something positive was coming to the city." Elena attended a debate by the city council of Kilkis, who explained how the ministry was doing this against their will.

These were people of the same parties who were in power in Athens! I understood that when mining companies claim that they are low cost gold producers – like Eldorado Gold does –

they do not mean the low costs for the environment and the people living in the neighborhood. They mean that they use methods that are much more toxic, to give the shareholders more profit.

Since then, it's Elena who's been faced with the incomprehension of people who do not yet know what a gold mine entails – and the prejudices towards activists in general. Her family is concerned. "My own mother once said: 'If you want to go to that demonstration and get injured or arrested, take your son. I will not take your role as a mother if you're not there anymore.'" However, Elena eventually managed to take her mother along for such a demonstration. "Afterwards, she told me she was scared but also proud on me!" In Greece, mothers and grandmothers are going to risky demonstrations. Not because activism is in their blood or because they enjoy it, but because they see the full scale of the threat and because they want to keep the community together. Their mother instinct wins it from their fear. Elena is often asked the question: is there an alternative? Her countryman Giorgos Velegrakis has an answer. He wrote a doctorate on the Halkidiki conflict, but until June 2015 he also served as a member of the Syriza party, in charge of writing an alternative development strategy for the area. His proposal contained four main points: cancellation of the project in Halkidiki, environmental restoration of the area, the gradual abolition of the "old" gold mining projects in the region and reconstruction of the local economy based on agriculture, ecotourism, forestry and fisheries. At present, that still seems a far way off. In 2018 the unfinished Skouries mine was still in "care and maintenance" with Eldorado stuck in legal cases and waiting for a government willing to bend the law.

Charlotte Christiaens of the Belgian non-profit organization CATAPA, which works with communities around the world on protection against mining impacts, is also convinced that the local economy can have a totally different base than gold: "Halkidiki is one of the most paradoxical places I ever visited: clear blue water, long sand beaches, rich fishing grounds, ancient forests, good goat farmers, beekeepers with the best honey, delicious olives...but also a toxic legacy of gold mining." What the Greeks especially need is not a new gold fever, but the right of self-determination. Through smart use of their renewable resources, the good life is within reach.

Undermining the middle class

Whether it's bauxite, uranium or gold: Congo has it all. What sets the Congolese frontline apart from the frontlines in India, Bulgaria and Greece is not the raw material – but the violence level of the struggle. In Congo, a terrible civil war caused over 4 million deaths. Claiming that this was a war over scarce resources is, however, a too simple explanation.

In March 2007 I travel to Katanga, in the southeast of what is very euphemistically called the Democratic Republic of Congo. My job is not just mapping the positions of the warring parties, but especially to map the motives of all those who are fighting. In its most literal sense.

A sort of minibus with wings and a front seat that passes for the cockpit drops me from Lubumbashi to a grass strip made in the jungle next to the village of Mitwaba. Without the flight, it would take weeks to reach the place. The next day I find myself standing right between a former Mai Mai warrior and a Congolese soldier. On other days they used to kill each other, but I'm in a rare safe place. The three of us are lining up for a plate of curry with lentils. The Indian UN Blue Helmets in Mitwaba are practicing curry diplomacy. All you need for that is a well-prepared aloe gobi, a generous portion of dhal and a mountain of basmati rice. According to the Indians, it's their *garam masala* that brings people together. The next day, a few of the Indian Blue Helmets bring me to a refugee camp, where the disarmed

Mai Mai try to survive. I soon understand why eating a curry is a reason for smoking the peace pipe. Between the undernourished mass of children, a woman tries to brew something in a cooking pot: leaves that are normally only for cattle.

In 2007, those Mai Mai still hiding in the surrounding forests had no good reputation. Some of their so-called wizards treat their fighters until they believe they are bulletproof. In a report from local human rights organizations, I read that, until 2003, this and their ritual cannibalism was used to scare off the opponent. The cover of their report is a picture where a crowd, including children, is looking at a pile of flesh. At the top of the pile there's a stick with a head pierced on it.[22] The still-fighting Mai Mai carried out sporadic raids on villages, after which they retreated into the dense jungle. The army controlled the villages and especially the mines.

The soldiers of the Congolese army were often not paid for months in a row. Many thus provided for their needs by using their gun to exhort all kinds of imaginary tolls and taxes from local traders. But their chiefs also encourage them to do that, because they want a piece of it. After all, they need to pay their superiors as well. If the commander of an army unit doesn't collect enough money for his superior, he's punished by being sent to a less lucrative area. That's called "reporting" and it's making money throughout the Congo flow from the bottom to the top, with AK47s as a lubricant. Police and military do not protect the people, they plunder the people. The state itself is an empty box: the structures and papers are there, but not the idea that the state exists to protect and help people. In theory, the army has the task of securing the mines. In practice, our maps showed that the army focused mainly on mines from which they can make more money, even if there are no rebels in the area. The really dirty work in the mines is done by the "artisan miners". That's a euphemism for kids as young as four who break their backs with heavy physical labor every day, for a starvation wage. Who you don't meet here in Katanga are the people who get filthy rich from all this. These are, for example, the fat cats of the diamond industry of Antwerp – the same people who go tell our government what tax rates they find okay to pay and even then evade those by going to Panama. They get to their Congolese source of wealth through a shady network of brokers in Rwanda and Congo.

Joseph Stiglitz, a Nobel Memorial Economics Prize winner, told on November 9, 2007, in a conversation with Gie Goris at Belgium's biggest book fair, that developing countries with plenty of resources and poverty need no foreign aid. According to Stiglitz, they need special help to get the full price for their raw materials. Because most Congolese wealth does not reach the state treasury, the Congolese state remains an empty box at best, but more a parasite on its own people in reality.

The research done by Steven Spittaels and this author resulted in a series of maps and the report *Mapping interests in conflict areas: Katanga*, which we wrote during our service at the International Peace Information Service, commissioned by the Ministry of Foreign Affairs.[23] We rejected the all too simplistic idea that all actors are merely fighting to enrich themselves with the riches of the earth. Our maps showed a different story. After many years of war, Mai Mai lost the courage and power to attack large mines, but they could no longer return to their villages, after their terrible way of recruiting and their many attacks on the civilian population. Their positions showed that. They were not at all strategically important places for commodities, but areas that are difficult to access for the army. They hid in forests, on steep slopes and far from the roads, where the army is at a disadvantage. Their battle in the years 2003 to 2007 was one of despair and survival, consisting of raids in villages and shelters in hideouts. But our research also showed that the Congolese army itself was a major threat to the local population.

These motifs are not cut into stone, they evolve over time.

For example, certain Mai Mai groups in the run-up to the 2012 elections suddenly appeared to commit attacks in the regional capital Lubumbashi, which indicates political motives. The struggle by everyone to enrich themselves with commodities is in any case a too simplistic representation of what happens in Katanga. Like in India, Bulgaria and Greece, there is a large group of people who lose wealth and well-being due to the mineral resources. This pattern is so prevalent that researchers call it the resource curse: the paradox that countries with many natural resources, especially minerals and fuels, have less economic growth, less democracy and less development. Academics have been discussing this paradox for at least half a century. My personal interest in this matter is the cross-border connections that appear to have a major and, in my view, defining influence on the local situation in India, Bulgaria, Greece and Congo. Another issue that fascinates me are the parallels in the broadest sense, between the war in Katanga and the world in general.

Is Katanga in a sense not the utopia for the most purist capitalists? The ingredients are the following: a shell state with no rule of law but where those with most money and guns rule. Where a desperate class does all the dirty work for close to no money. There is almost no big and well-known company with activities on the frontline itself, yet there is a lot to be gained behind the scenes. And if a serious journalist exposes shady deals, there are still lawyers to pay to sue them, as happened with MO* magazine.[24]

On dark days, I sometimes fear that humanity evolves towards this kind of deregulated capitalist utopia, which is of course a dystopia for the vast majority as well as an ecological nightmare. It's a state of consciously maintained chaos and survival with extreme contrasts between rich and poor. In 1994, Robert Kaplan wrote the groundbreaking article *The Coming Anarchy*, which explains how, among other things, scarcity and overpopulation will give rise to an all-around anarchy of violent nature.[25] He said that environmental problems and scarcity of natural resources will be THE security issues for the beginning of the twenty-first century. I do find Kaplan way too mild for the influence of mighty companies, institutions and political leaders in the West but he is good at connecting a certain geographical reality with associated deeper political and economic conflicts.

Are the contours of a Congolese dystopia recognizable at the global scale? Maybe some elements are already upcoming. You have a class of people banned from their birthplace and system. In Katanga this is the Mai Mai. In India, it's the dozens of millions of people who had to make way for a reservoir, factory or mine. But even in England, the number of people on the street doubled between 2010 and 2015. For them, there is no place in our society. They survive by hiding in the crusts of the earth; from the forested slopes in Katanga to the slums of Mumbai and the train stations of Europe. They're in a struggle to survive, which often forces them into begging or sporadic raids on society. Not to get rich, but to survive. This wreckage in a turbulent ocean is faced with coastlines that consist of ever higher walls. In the slums of society, you are anonymous, stateless and powerless. This class is unfortunately booming.

Professor Saskia Sassen describes in Expulsions how in the last 2 decades a fast-growing number of people have arrived in that class. It's not a natural process. She connects dots like the removal of the unemployed from the social welfare state and the rise of mining techniques that remove life from the biosphere to the number of people who are on the run. Between 2005 and 2015, the number of people fleeing war doubled, to well over 60 million. By 2016, a record number of 65 million refugees couldn't go home. To them you have to add the 15 million people who are displaced each year for the purpose of large-scale development.[26] This displaced class grows further each year, fueled by the decommissioning of the social welfare state and by companies that put profit before people and that are faced with ever more big-business supportive states.

Above this "lowest" class of "stateless people" you have in the East Congo the subclass that works for the state. In the few months Kinshasa has not forgotten them, they get some starvation wages. But they do have weapons and stamps that enable them to rob people of money. That money is necessary to make their patrons happy, but they are getting some comfort prices. These Congolese soldiers – but also many workers and servants who have the luxury of a more legitimate means of income – do that in return for that little bit of security, a minimalist form of control and freedom. They do that in service of people above them who claim a growing part of the pie. Their subordinate role behind the PC, factory band or AK47 is frustrating but a strike here and an armed revolution there allow for the occasional sense of ownership over one's own destiny. In the past, strikes delivered many useful things in the West, such as paid holidays, less working hours and voting rights. They helped to create what we now know as the workers class. But far more power is in the hands of people above this group – which too often talks about the workers class as parasitic, greedy, lazy and too stupid or too weak to gain the kind of wealth they enjoy. And because that small group above the workers class also pays most of the media, we hear that a lot. This is the class of lieutenants, the lower boardroom and the better off self-employed people who've made it and have hired a handful or more people. Over the past few decades, this upper class has been able to see their profits rise faster than the workers class. They've often gathered enough money to let the money work for them and make them even richer while sleeping – thus breaking free from the trap that most people below them are in. These 2 or 10 or 30 percent (depending on the region) saw their share of the cake in the last 40 years of increasingly deregulated capitalism increase exponentially. I would like to refer the lover of more precise figures to Thomas Piketty's highly-honored reference work: *Capital in the twenty-first century.*[27] Within this group you have people who sometimes work twice as much as the average, but get ten times as much as the average. Often they regard themselves as the essence of the economy, and think that if it's going well with them, it's going to be good with the economy and with everyone. These myths spread through media and political friends.

And then you still have that real 1 of often just 0.1 percent at the top. The hedge fund shareholders and CEOs who are simply not interested in the blood sticking to their diamonds because it's way too far removed from their world. These are the fat cats that have earned the same amount on January 2 which we in the workers class work for the whole year. They raced ahead even faster. A CEO of one of the 500 largest US companies now earns 204 times more than the payroll member in his company. Half a century before that was 20 times more.[28] And the trend is not limited to the US. This global jet set live in a bubble that takes more and more space.

The people in that bubble rarely see that the number of needles around them are increasing. But some do see it. One billionaire – who calls himself an "unapologetic capitalist" – wrote a long article in which he warns his friends from the plutocrat class. The title of the article was "The pitchforks are coming...for us plutocrats".[29] If we continue with capitalism, that is not unthinkable.

Of course, I admit all this is a rudimentary division. In reality, a much greater degree of complexity and regional differences exists. But the class issue exists, cannot be ignored and the connections with the way we extract resources to meet the ever growing demands at the top are a death or life issue for millions and maybe even billions of people. It's a conscious choice to take the "middle class" out of this rudimentary division. Professor Saskia Sassen concluded after her investigation that there is a difference between capitalism a few decades ago and capitalism now: capitalism no longer creates but destroys a prosperous middle class. Professor Jonathan Holslag is also sure: "The middle class, she's dying."[30]

Europe's middle class is eroding, especially in Germany.[31] The International Labour Organization claims this is due to privatization. In countries with major austerity programs, such as Spain and Greece, the middle class melted like snow before the sun. In the US, she's been on the way back for decades. The choice is simple: either we make a political U-turn on the causes of this great decline of the middle class or we keep moving in the direction of the class divisions in Eastern Congo – a kind of capitalist Walhalla with no middle class. I'll be blunt on the latter option: you don't wanna go down that route (unless you're a dirty diamonds trader).

The good news is that even in the current period of degradation there are also signs of an upcoming major turnaround. The number of people pulling the emergency brake on the austerity train increases. With Trump as US president it's easy to forget how close the socialist Bernie Sanders was to the presidency. In the Netherlands, Belgium, France, Iceland, England and all over southern Europe the people who propose a systemic change in progressive direction are surfing on a momentum, despite the enormous pushback by the better funded systemic players.

Less visible than those political alternatives – which we sure need – are the many groups of people who are building the economy of the commons. This economy is mostly detached from both the free market and the state. Once you understand that the self-destructive extractive economy is doomed to collapse – even though it is not clear when and how – you might want to prepare for that in two ways. First: build up an extensive back-up system for yourself, your family and your community, which can keep going once the GDP economy collapses. Second: keep working on a larger policy alternative and make sure as many countries as possible already test, develop and refine it as often as possible. A political plan B in which a majority can thrive without slowly destroying the air, water and soil on which all life on earth depends. More on these parallel pathways that make a U-turn and lead us away from the edge in the last part of this book. First, we continue the journey along the many frontlines that capitalism keeps opening up. Bauxite, uranium and gold are far removed from most people's daily reality, or hidden in some aluminum frame, an electric wire or your phone. But we all know and see sand. Let's travel from (perhaps) the poorest place on earth to (probably) the richest: Dubai.

Sand: one big Fata Morgana

In 1995, I leapfrogged the then still-powerful middle class, vaulting from relatively poor (but not desperate) directly to rather filthy rich. For a full 2 weeks that is. A much wealthier friend had moved to Dubai and, as a 15-year-old teenager, I was invited there to spend the Easter holidays. One day his kind mother took us to a five-star hotel, private beach included. We played golf in mind-bogglingly lush greenery in the middle of the desert. If you, like me, associate the word "holiday" with a nearby campsite, Dubai is a fairytale world.

The potential for fun in Dubai is endless. The only other competitor for infinity status is the sand. A 4-wheeler took us to a region where your life could depend on not getting a flat tire: the Empty Quarter. The capital E and Q are well deserved. Think Lawrence of Arabia, camels and a thirsty death. This is the largest continuous sand desert in the world. The English explorer Wilfried Thesiger described the Bedouins roaming the desert as a hospitable, courageous people of amazing endurance. That was just as well, because the Empty Quarter has been one of the hottest, driest and most inhospitable places on earth for thousands of years. Sand dune after sand dune stretch over an area larger than France.

Today, you do not need Bedouin blood to traverse the

landscape. An air-conditioned jeep springs across dunes like an Arab charger. In Dubai everything lulls you into believing that nature can be controlled at the touch of a button. Whether it is the speed and stability of a jeep zooming across the dunes, the ubiquitous air conditioning or the towers that seem to defy the laws of gravity. The scorching sandbox has been reduced to decor on a movie set and we are the stars of the picture.

Returning to Dubai in 2013 by plane, I glimpse a mile-long artificial peninsula created in the form of a palm tree extending way out to sea. The strip is thick with hotels. A little further down the coast and a sort of world map has been made out of artificial islands. They were slowly sinking into the sea. When the global recession of 2008 came to Dubai, the world stood still. Well, the project with that name experienced a sudden lack of funding as the taps of global finance ran dry. By then, ships had already spurted 321 million tonnes of sand into place.

Elsewhere, construction continued. Dubai now has the Burj Khalifa, the highest tower in the world. Standing at 828 meters, it offers neck pain and dizziness just looking to its heights from ground level. How much concrete does it contain? About 330,000 cubic meters according to the building's website. A major component of the concrete is sand, which is easy peasy to score, what with so much of it lying around.

But as it turns out, the sand in the Burj Khalifa all came from Australia. That is because there is not enough sand for concrete available in the region.[32] Unimaginable as it may seem, the fact is that the largest continuous sand desert in the world is unsuitable for concrete. It is not even good enough to build those palm islands. The wind has free rein in the desert and has made sand grains too round, so much so they do not stick together. Not all sand was created equal. Think of trying to make a sand castle with the dry sand where the high tide never reaches. If you have ever been a kid on a beach or a parent that takes beach fun seriously, you know that does not work.

Marine sand is much better for the job, but the lion's share of the local stuff has already been used up for the palm islands. What is more, the salt in dredged sand is a bad match for steel in reinforced concrete. Dubai desalinates its drinking water, but even with its fabulous wealth, this would be way too costly for cleaning marine sand. It also requires oil, and unfortunately for Dubai, its oil stock is dwindling.[33] The city already imports more petroleum products than it exports and in a decade or two its wells will run dry.[34]

The potential for a collapse of Dubai is kept well-hidden as the city radiates self-confidence. In many ways, this seems just a concentrated case of a more global discrepancy. Host of The World Expo in 2020, Dubai will probably put up one of the world's most profligate shows. A tower even higher than the Burj Khalifa is being built to be ready by the time of the event. In 2012, Barclays Capital subverted the popular saying "pride comes before a fall" with a study pointing out that "high-rises come before a fall".[35] The study says there is a strong chance of financial crashes following a boom in the construction of skyscrapers. If you look past the palaces in Dubai and its sinking oil, water and good sand reserves, the question seems to be not whether, but when, the desert will blast Dubai's bling into a set suitable for an apocalyptic film.

Nearly 6000 kilometers east of Dubai is Singapore, a nation that stockpiles sand.[36] It imports massive amounts and keeps it as a reserve, comparable to a strategic stock of oil. Singapore needs sand to literally continue to grow. The city-state has grown its landmass by a fifth in the last half century.

Initially, this was an easy task, as Singapore's neighbors were willing to sell their sand. Things changed in 1997 when Malaysia stopped the trade. Indonesia and Cambodia followed suit in 2007, Vietnam in 2009. The entire international sand business became a political minefield. Singapore was growing at the cost of destroyed beaches and rivers in other countries. Populations of these countries tend to dislike the mere idea of selling pieces of their country for the purpose of expanding another country, especially if violence against them and their environment is involved. In some cases, the export went underground, no pun intended. Global Witness found that in Cambodia, the most corrupt country in South East Asia according to anti-corruption watchdog Transparency International, officials continued pushing export contracts worth millions.[37] Investigative reporting has shown that a very similar scam was going on in Vietnam, also for export to Singapore.[38]

Sand tends to be extracted from vulnerable natural areas, a process that effectively stops local fishermen from reaching their most important resource, fish. They are forced to avoid the troubled waters that the sand-mining in rivers creates.

The sand mafia is capable of staggering feats of engineering. In Indonesia, 24 entire islands were sucked up and removed from the map in order to sell sand to Singapore.[39] The disappearance caused a bizarre dispute over the exact location of the international border between Singapore and Indonesia.[40] So hungry is Singapore for sand that it was prepared to pay $0.19 per kilo. To put that into perspective: In January 2016 the price for one liter of crude oil went down to $0.17 per liter. In Singapore, sand is the new oil.

The sand mafia in India

Scarce but valuable resources attract conflicts and as with oil, some of them turn violent. Vince Beiser, a journalist and author with a solid track record of covering the sand issue, wrote that battles waged by sand mafias have killed hundreds of people in recent years in India alone, "including police officers, government officials, and ordinary people".[41] Through a contact in India I find someone who has so far survived the locking of horns with this mafia.

In Bollywood's backyard lives an activist that even blockbuster movie writers would struggle to invent. Sumaira Abdulali, or the minister of noise, as the media have dubbed her, is a 55-year-old public figure in India. She won the title for her activism against noise pollution in the city.[42] But she actually won her spurs fighting the sand mafia. In 2004, she noticed that the beach near her house in Alibaug, near Mumbai, was shrinking. She heard trucks at night that she suspected were carting the sand away and decided to take action. She called the police, took her car and drove to where the road ended at the beach, expecting to meet them there. "Instead of rushing to the crime scene, the police apparently warned [off] the illegal sand miners," said Abdulali. As she waited in her car for the police to arrive, the men present at the beach pulled her out and assaulted her.[43] She survived, but was hospitalized As she was beaten, a man from the mob asked: "Do you know who I am?" His father was the owner of a construction materials company with a near-monopoly in the area and an important local politician. The point he hammered down was: don't mess with us.

But that's exactly what Abdulali did. Two years later, she started a lawsuit. With success. In 2010, the Bombay High Court banned sand extraction, a ban which remained in place until 2015.[44] Abdulali tots up the cost of sand extraction in India: soil erosion, landslides, falling water tables, infertility of farmland, disturbances of ecosystems and marine life, beach disappearances and collapsing infrastructure. Once Abdulali made a surreal video of a train crossing the Vaitarna railway bridge while machines were illegally extracting sand from the nearby riverbed.[45] This imposing colonial-era bridge is in Virar, north of Mumbai, and remains the city's sole rail link to north India. Railway officials went on record with concerns that the Vaitarna bridge's foundations had been weakened by sand-mining.[46] A senior Western Railways officer told The Hindu newspaper: "We believe there is a nexus between the sand mafia and certain state government departments. Due to the illegal sand-mining, the flow of the Vaitarna has been altered, which is a dangerous sign for the bridge's health."[47] In August 2016 the Mahad bridge across the Savitri river collapsed, killing at least 28 people.[48] Several activists, including Abdulali, blamed the incident on sand-mining of the riverbed.[49] Sand excavating equipment was discovered under the bridge, which is in close proximity to the attack by the sand mafia on her in 2010. In spite of this, the Government, without conducting any investigations, dismissed the possibility that sand-mining may have had anything to do with the collapse of the bridge. Even now, mining continues in the area.

Abdulali didn't stay in splendid isolation. Other courageous people have started exposing a dirty secret. In early 2017, journalist Sandhya Ravishankar wrote a four-part series on illegal sand-mining along Tamil Nadu's southern beaches for the online magazine The Wire.[50] She has since received death threats and online abuse.[51] Ravishankar is in no doubt about whom she believes is behind the harassment – the sand-mining company named in her reports. In March, she filed a complaint with the Chennai police.[52] In India, anti-sand mining activists are often attacked and even killed.[53] Journalists are not safe either.

In 2010, Abdulali took a journalist and photographer to visit Raigad, a Maharashtra district where sand extraction was in progress, despite the ban.[54] The three posed as real estate agents and began filming the illegal extraction taking place on an industrial scale. Their cover was soon blown and what followed was a nightmare return to Mumbai that they were lucky to survive. Their returning car was pursued by two cars that tried to ram them into a ravine. As they sped back down a dirt trail, they knew that overseeing any of the many potholes would be fatal. At the main road, a truck was waiting. While crossing a bridge the driver tried to overtake and bump them in the river but fortunately he didn't get further than her bumper.

"What saved our lives that day is that my husband is a professional rally driver and he taught me some of his driving skills," said Abdulali. "Again, the police were in cahoots with the mafia. I wanted to report this murder attempt, but they wanted to give me a speeding ticket!" Once again, a powerful local politician controlled the illegal trade in sand.

The Bombay High Court later severely criticized the police for how they dealt with the attack on Abdulali and her companions.[55] "India's sand mafia usually co-opt the local village council or higher political figures", Abdulali says. One local sand miner went on to become a minister of state in Maharashtra. His portfolio? Environment. The fox was well and truly put in charge of the hen house. Abdulali: "He claims he is out of the sand business now, but still owns the largest sand storage site in the state. He has shifted his business interests to the next construction industry cash cow: stones." When asked for solutions, Abdulali says the recycling potential in India's construction sector is by and large untapped.

To reduce demand for new sand, you need to evolve into a circular economy. Big cities in India crush many old buildings to make room for the new, but the debris ends up at landfills. In some countries, the use of primary material is only allowed after the demolition waste is used up. In the Netherlands, 90 percent of all demolition waste is recycled. Even poorer countries like Vietnam are now reusing demolition waste. You can build roads with a lot less sand by recycling plastic as a resource.[56] We have to do that. If we continue like this, India will dig a grave for itself and pay a very high price. The circular economy is a much better option.

* * *

A circular economy creates some novel scenarios. One beer company now claims you can drink (their) beer for the sake of saving beaches.[57] Empty beer bottles can be turned into a kind of sand that is useful in construction. Of course, a circular economy also needs energy and if the energy used in the circular loop is of fossil nature, questions need to be raised whether it is part of the solution or part of the problem. But the beer company does have a point. Beaches worldwide are in trouble. Beaches are made by rivers gradually shifting sand to the sea. Take the sand out of the river and you end up without a beach. In Sri Lanka, they found that out the hard way. The country's most eroded coastline is around the delta of the Maha Oya river, which is where most sand-mining takes place. In some places the beach recedes by 12 to 15 meters each year.[58] Thousands of families here have seen their land erode into the sea. When the damage became too rampant and too evident to be ignored, a ban on mechanical sand extraction was issued. But here too, that did not stop the sand mafia from continuing to dig deeper.

Sand mafias exist not only in India, Sri Lanka or Asia in general. In Elmina Bay in Ghana, it even digs sand just in front of the few beach resorts the country has. Hotels have lost 30 meters of beach. The sea now laps at their doorsteps.[59] The Environmental Justice Atlas charts at least 71 conflicts triggered by sand, gravel and quarrying. They exist in all continents. And so I wondered: has a sand conflict arrived at my doorstep as well?

Flanders digs a grave for itself

Dubai, Singapore, India, Cambodia, Sri Lanka, Ghana: so far it may sound like sand supply is an exotic problem. Belgium has just 60 kilometers of coastline, but millions of people visit it every year. On any given weekend with temperatures above 25°C, hundreds of thousands scramble to secure a spot on the small strip of extremely valuable, micro-managed beach. If there is one beach that should be doing just fine, it must be ours, right? Just to be sure, I checked with an expert in our coastal ecosystem:

Katrien Van der Biest. We studied geography together and shared a passion for observing the forces of nature up close. The last time we stood on a Belgian beach together we were skipping class in order to witness a fierce storm making landfall. Katrien honed her passion towards oceanography and a doctorate on the many benefits of healthy coastal ecosystems, not least their many sand dunes and banks.

Digging holes in densely populated Belgium is rarely a great option. Therefore, from the mid-1970s onwards, extraction started to move offshore. In the first 10 years, machines dug about half a million tonnes a year from our limited zone of the North Sea. Now it vacuums up 3-5 million cubic meters a year and the next 35 million cubic meters are already scheduled.[60] Katrien says that sand is a finite raw material; that North Sea stocks are shrinking and that the industry is aware of it. Meanwhile, parts of sand banks close to the coast are now closed for mining after reaching their legal capacity. They are not recovering as hoped, so the current trend is to move to sandbanks further out to sea; the next frontier. But behind that wait the sand-poor international waters and the end of story for sand-mining in Belgium.

The industry is sucking up hidden treasures, according to Katrien. She describes how they provide invaluable services, such as breaking the back of oncoming waves that might otherwise cause problems on land. A lot of North Sea sand is used to reinforce beaches and provide coastal protection in general. That is until the next storm sweeps it back out to sea. Demand for this application has been rising rapidly since the mid-2000s and shows no sign of slowing. As in Asia, treated North Sea sand is also used by the construction industry; specifically coarse sand, a much scarcer resource than fine sand. Legal limits placed on extraction are currently being debated. The sand tzars are, of course, hungry for more.

Vera Van Lancker, senior researcher at the Royal Belgian Institute of Natural Sciences, emphasizes that Belgium is a global leader in mapping marine sand, a precondition for managing it sustainably. A well-respected expert on the subject, she explains that: "if the government adjusts the 5-meter limit [on extraction], it will do so because we now understand better which kind of extraction causes more or less damage". But she too warns that: "In the long run, we're looking at a shortage of sand and we'll probably have to import it from the Netherlands."

So we are currently going by the assumption that once we have exhausted all legal sand stocks in Belgium, we will import it from our neighbors But is that not a reckless assumption? The Netherlands recently announced it will stop selling gas to Belgium no later than by 2035 because its stocks are running out and it wants to use what little remains for itself. While the Netherlands has much more sand than Belgium, they too have experienced an exponential rise in sand-mining and with a rising sea, they're going to need a whole lot of it in the future. Van Lancker also highlighted new research showing that sandbanks do not recover as well as previously thought. The stocks of the more valuable coarse sand in particular are declining. "It all shows that we need to take more circular economy measures, while fully realizing that in the end, that alone will not be enough," Van Lancker said.

The man responsible for tackling some hard questions on the topic, Belgian state secretary for the North Sea, Philippe de Backer, says that for him, the economic potential of sand extraction is the top priority. "We have to make use of the economic potential of the North Sea without losing sight of the ecological importance," said de Backer. But when de Backer speaks of "ecological importance", he means the local marine environment, not the impact of sand-mining in the sea on the coastline. He points out that he is not responsible for coastal defense and refers questions on this to his Flemish counterpart. Welcome to Belgian surrealism: one governance level of the federalized state shamelessly digs a grave for the other.

Of course, de Backer does have a point: it is in Belgium's short-term economic interest to go for growth, meaning more extraction from the North Sea. After all, Belgian companies Jan De Nul and DEME are global players, taking numbers two and three in the worldwide dredging sector. A few of my former classmates now work for them, but none are willing to speak about sand scarcity or their employer's take on the issue. With shortages forecast, heads in the sand seems to be the order of the day.

Jan de Nul's PR department gives me short shrift: they want nothing to do with this book. I found just one insider willing to give me some background. We talked at length but when asked if I can use some of it for this book my source refused. "It's a small world and if you give details, they'll know it's me who leaked," the source said, unwilling to go further. Some journalists did manage to find some dirt, though, for example about Jan de Nul's record in Indonesia, where the sector has a rotten reputation.[61]

Meanwhile, a study from back in 2008 shows that widening the navigation channel to Oostende harbor to allow bigger ships has seen waves grow by 10 percent during storms.[62] That was before the extraction industry upped its take from 0.5 to more than 5 million cubic meters of sand per year. This weakening of our natural storm defenses is coming as sea levels are rising exponentially and storms are getting more frequent and extreme. Just when we need more coastal protection, we are removing them for short-term and very private gain.

Neither the so-called free market nor the Belgian government seem willing or able to see this perfect storm brewing, let alone deal with it. It is a textbook example of massive future public costs for current private gains. All this happens despite a major warning. In 1987, the Herald of Free Enterprise sank shortly after leaving the port of Zeebrugge. The captain was in a hurry and motored out to sea with the ferry's rear doors wide open.

A court later concluded there was a "disease of sloppiness" and negligence at every level of the corporation's hierarchy, with vital equipment not installed in order to cut costs. The Herald of Free Enterprise sank in 90 seconds and took 193 people to the bottom of the sea.

Today, we are facing a kind of mega version of the Herald of Free Enterprise. After all, the free enterprise economy is much more mature now than back then. Powerful dredging companies are currently opening the back doors of Belgium. Often on behalf of the government. It will take longer this time, but there are a few hundred thousand Belgians aboard.

Some people see the perfect storm coming. Flanders has a spatial planner. His job is to figure out what Flanders could look like in 2100. He hedged his bets somewhat by coming up with four scenarios for the Belgian coastline. Two of these are sure to give homeowners a sinking feeling. A new dike starting at the midway point of the 60 km long coastline would go straight into what is now inland while the sea engulfs the Flanders west of it in a major but planned way.[63] The whole western half of the coast and the fertile low-lying polders in its hinterland would be underwater by 2100. In the report, this is framed as "active re-wilding" and "innovating" but one can doubt whether the hundreds of thousands of people who now live there will ever make that terminology theirs.

Now, I can explain to my 8-year-old daughter that if she wants seawater to flow into the moat around her sandcastle at high tide, she needs to dig a big gully directly towards the sea so that the waves wash inland faster. Every year we enjoy this classic beach holiday ritual. But the spades of Jan De Nul and co are slightly bigger than ours, and the castle is West Flanders.

I asked Katrien if there is something wrong with my childish logic. I want to know whether there are any scientific studies that substantiate the parallel between gullies dug in front of a sand castle to get the water more inland and the extraction of sand banks as a way to get the sea more inland in a big way. Katrien responded, "We know there's an impact on the environment, but the scientific evidence of the precise effects on ecosystem services such as coastal security is still lacking." That is odd for a state that endorsed the UN precautionary principle back in 1992.[64] It is also odd for a state that is the biggest supplier of marine sand.

It only makes sense when you apply a capitalistic logic, that a massive cost in the future should be discounted and thus not included in the price of today's sand. The assumption is that we will keep adding GDP so fast that these future costs will be peanuts compared to the size of the economy we will have in 2100. Another option would be to look at the whole ecosystem and think about the coming and going of the sand itself. A state that aims to protect its people would want to ensure that in these times of climate change, the net balance of the protective sand banks guarding the coast is positive, not negative. The only thing standing in the way is the dominant political climate. It is the die-hard belief in the invisible hand of the free market, even if that hand is busy sabotaging the floodgates.

The coming sand wars

I found no hint of a sand mafia in Belgium. But even a so-called leader in sand stock management, like the Belgian state, is facing a dire future. There are many more local stories to tell, even within Belgium, but I want to zoom outward to the bigger picture.

Worldwide, we use twice as much sand as is transported in all the rivers throughout the entire world. Our insatiable hunger for sand has seen us start to dig elsewhere. The majority of all the sand we now use is marine sand, and as a result, two-thirds of all beaches in the world are in the process of either retreating or disappearing. In northwest Europe, more than 100 million cubic meters of marine sediment is extracted from the North East

Atlantic, mainly sand from the shallow North Sea.[65] Because this marine sand is less suitable for concrete, it needs to be washed with fresh water. That poses another problem.

Fred Pearce, the acclaimed author of *When the Rivers Run Dry: Water—The Defining Crisis of the Twenty-first Century*, points out that if everyone lived like the average meat, beer and milk consuming Westerner, all the water in all the rivers in the whole world would not be enough. Forget the one or two liters of water you drink every day. Making an average scoop of ice cream uses up to 1000 liters of water. One steak swallows up 5000 liters.

The world's soils provide twice as much food today as they did a generation ago. But in that period we have also diverted three times as much water from rivers and lakes to agriculture. At some point, choices will need to be made between using fresh water for food crops or for washing sand from the sea. In June 2017, something remarkable happened in rainy Belgium. After a dry streak, the government issued a temporary ban on using water for agriculture in West-Vlaanderen, Belgium's coastal province. Competition with the water demand from its massive tourism sector is fierce there. The interests of the construction industry, the tourism industry and those of farmers are clashing. We have come to the point where even in Belgium, where it rains almost equally the whole year round, water is occasionally scarce.

The choice between the beaches and buildings is also a hard one. The US opted for beaches some time ago, at least when it comes to preserving its own. A federal ban was put on sand-mining of beaches to keep them intact. Yet there is still one company openly digging away on a beach in California thanks to an obscure legal loophole it is exploiting. The construction company that is literally making America smaller again has its headquarters in Mexico.[66]

This cheeky jab in the eye would surely drive President Trump to a flurry of Twitter invective. But the big man has other conflicts

access to oil and gas reserves. Tensions between China and other countries around the South China Sea have increased in recent years, sometimes resulting in clashes and fatalities, such as the conflict between China and Vietnam. After China acted in a way that Vietnam considered to be an aggression in its territorial waters, some mobs killed Chinese people based in Vietnam. Still, the much bigger issue is the rising tension between China and the US. The Trump administration does not seem very willing to accommodate China as the superpower it has become. Trump's former strategist and right-hand man Steve Bannon once said that the US and China will be at war within 5 to 10 years, and that such a war would begin in the South China Sea.[70] Trump's former secretary of state Rex Tillerson, formerly the CEO of oil giant Exxon, claims that China should not be allowed access to the islands it is building in that sea. This prompted China to remind Washington that the US was not a party to the conflict in the South China Sea and it would be wise to keep it that way.[71]

All of this is uncomfortably reminiscent of a classic work of science fiction from the 1930s. In that era of emerging fascism, Nobel Prize-nominated author Karel Capek wrote War with the Newts.[72] The book satirized the shortsightedness of the then prevailing appeasement towards an essentially evil and destructive regime – the one led by Hitler. Capek wrote about the erosion of vigilance towards a growing evil which in the end led to the erosion of the world's coastline and the collapse of human society – by the very same Newts that the evil regime had first created to work for them. He used prose, a metaphor and a rich imagination to illustrate how a raw imperialist expansion strategy quite literally eats itself up from the inside out. In other words: karma is a bitch. Little could Capek know that one century down the road, his absurd fiction was in a way coming true in real life. Humanity has created monsters that are eating our coastlines and rivers, that are bringing us ever closer to a collapse of civilizations and that are doing this as part of their function in a self-destructive system: capitalism. The Newts are simply replaced by the sand mafia.

Chapter 2

Fossil fuel

Chevron's "Shock and Awe" doctrine

October 2014. I call an Ecuadorian friend visiting The Hague. "We are about to walk into the International Criminal Court. I'll call you back when we're ready." He hangs up, steps in and leaves an explosive package. Forty minutes later he calls again. "We did it!"

Julio Prieto had struck a direct hit with a bombshell that could end up doing untold damage to one of the world's largest oil firms. He penned the report of 50 explosive pages from the heavily polluted jungle of northeastern Ecuador. An area baptized by locals as "the Chernobyl of the Amazon". Decades of oil extraction by Texaco (bought in 2000 by Chevron) left a much bigger explosive package than Julio ever could: somewhere between 16 and 20 billion liters of highly toxic "oil-mud" laying waste to an area larger than Luxembourg. The Grand Duchy of Luxembourg, not the city. The death toll of that poison has reached over 10,000 souls and counting. As it continues to seep into water courses and food chains, Texaco/Chevron's insidious chemical warfare will continue to claim the lives of subsistence farmers and indigenous forest people. This will continue until a major cleaning operation is carried out, the kind of clean-up Ecuador's supreme court has ordered Chevron to pay for. But Chevron's CEO, John Watson, prefers to spend a few billion dollars on lawyers, PR firms and spies rather than on saving the lives of thousands of victims. This makes him – according to prosecution lawyers – personally and criminally responsible for all the body count that has continued rising ever since Chevron decided to stonewall the historic 2013 ruling. If Julio and co ever succeed in nailing Watson, it would have enormous implications for the oil industry and beyond. Judges in The Hague are not known for breaking speed records in delivering justice but they signaled they are opening up to environmental crimes and they might take this case on as a test-case. So while the 50 pages are only the metaphorical bomb, there are good reasons to give their potential to jail the Chevron CEO a superlative adjective.

I spoke with a determined Julio Prieto in Brussels, the day before he went to deliver his explosive package in The Hague. The soft-spoken lawyer represents the *Afectados,* a band of 30,000 affected Ecuadoreans fighting to make the polluters clean up the mess they left behind. Julio is the right-hand man of the charismatic Pablo Fajardo, who leads the legal efforts, and together they face one of the giants in the oil industry. Julio makes a humble impression for a lawyer, preferring jeans and a jacket over a sharp suit.

I help with the legal actions abroad. The victims opened a lawsuit in New York in 1993, but Chevron wanted to be judged in Ecuador. After many years of legal battle, they obtained this right. However, the New York judge added that the condition for that is that Chevron would respect the outcome in Ecuador. By 2013, the supreme court in Ecuador finally decided that Chevron has to pay $9.5 billion plus interests for cleaning up the mess they left behind in Ecuador. Chevron refuses to pay. So, we convinced lawyers in Brazil, Argentina and Canada, where Chevron still has enough assets, to enforce the legal decision in those countries.

Legally speaking, the case is over. Chevron lost. But what does the poisoning of a habitat affecting tens of thousands actually mean? A sense of the carnage emerges from the 220,000 pages of court papers; more than 100 expert reports, plus the reports of 54 independent field inspections ordered by the court. The oil operator was found guilty of using substandard techniques to store the waste from the oil extraction. This saved them money, but caused massive leaks. The court demanded $5.4 billion be paid to clean up heavily polluted soils, $1.4 billion to meet health costs, $600 million for research and eventual treatment of groundwater pollution, $200 million for ecological restorations other than soil and water, and $150 million for a drinking water system. A paltry $100 million was reserved by the court as compensation for the 30,000 victims.

Nine billion dollars – actually more than 12 billion today due to interest – is a lot of money. But not to Chevron, which rakes this in every 20 days. In fact it took the firm just a few days to earn the $2 billion it diverted into a fightback campaign against the *Afectados*. This dirty war was prosecuted by a 2000-strong legion of lawyers, PR ringers and spies. The aim was simple: delegitimize the opposition and turn the tables on who's the victim in this case. The offensive became perhaps the most comprehensive persecution of affected people and their lawyers in the oil industry's long and dirty history. Chevron deployed a legal strategy normally reserved for mafia groups and terrorists. It accused its opponents of being a financial extortion outfit and used the anti-mafia Racket Influenced and Corrupt Organizations laws to put a stranglehold on the opposition. Dozens of marginal players were hauled into court in an effort to choke off support for the *Afectados*. The company made headway when it convinced a judge in New York to convict opposing lawyers. The victory was a sham; the prosecution and verdict a travesty of the US legal system. It later transpired that the judge was a Chevron shareholder and more importantly that the company had spent $2 million bribing the most important witness to commit perjury – a fact he admitted after the ruling. Nevertheless, the *Afectados* legal team had been hamstrung. Julio:

They gave us the treatment that is normally given to terrorists. As a result, we had to dissolve our legal team of four, due to the lack of financial resources. But this will not stop us. Pablo, I and many volunteers will continue this until the end. Before all this we were not wealthy lawyers either. Many of my former classmates work in business law. I'm that one romantic human rights lawyer they probably joke about. However, I have been doing this for 12 years and I would do it again. Only in the past 3 years, Chevron's counterattack made our life really difficult. It's hard to swallow your pride and, at the age of 34, ask your father for an advance on the rent of your flat. But my parents and friends support me. This trip to Europe, for example, is fully paid by donors. This is all a matter of time. There are three countries where we will eventually win: Canada, Brazil and the United States. And we only have to win one.

As our conversation becomes more informal we exchange stories of unpaid activism and how our partners, parents and friends react to our struggles. I want to know more about how Julio deals with internal struggles that are rarely seen or talked about. About his partner's reaction, he says: "In the past, she sometimes said, 'you are a lawyer, when will you find a real job?'" He laughs. "I just asked for patience. But now she really supports me. She was a business lawyer, but she also stopped her job and now she also works around human rights, finding it more rewarding in non-financial terms." The sterile setting of the room I had booked for the interview starts to annoy me. So I ask the Belgian Solidarity Committee volunteers with whom Julio travels if there is time to sneak out to the bar before he needs to move on to an evening lecture. A freshly tapped Belgian beer in one of Belgium's most hipster hangouts, café Belga, helps to learn more about the man behind the lawyer.

Julio's parents cannot have been poor, since they were able to pay for his studies at the best university of Ecuador's capital, Quito. His teacher was already acting as a lawyer for the *Afectados* and one day he was asked to join him. Julio: "I didn't think or doubt but just went. After a few years, my teacher had to bow out, but I stayed." In those first years, a mere career opportunity became a lifetime struggle. I ask how hard it is to go from the best law school in the capital to working with indigenous people in the jungle?

In the beginning, I just did the legal work in Quito. But eventually I began to go to the jungle and meet people. I began to listen to their story. If you listen long enough, you get involved. Once you see the pollution and know the sick people, there is no way back. In the jungle I also met Chevron. We arrived with a team of four in a small van, they arrived with 20 lorries. Their well-paid business lawyers screamed lies about Chevron here and Chevron there. It was utterly arrogant and disgusting.

I asked if the experience riled him.

Well, one week after setting up a local office, someone broke in. Of course, Chevron did not leave a note to say "we were here", but nothing was stolen, not even the easy to find money in a drawer of the desk. Then there were a couple of conversations with people on the street who asked us weird questions and others who took pictures of us all the time. We went to the Inter-American Commission for Human Rights and we received personal protection through the Ecuadorian state. But those escorts soon became a burden. Wherever we went, we had to pay their hotel and costs and we simply could not afford it. So after a while, we said "Thank you very much, but we will do it without you."

Julio showed me the sticker he had placed over the camera on his laptop and added:

If I could jam the microphone too, I would. Did you know that Chevron paid spies from Kroll to follow us? Kroll is one of the largest private intelligence services in the United States. Until 2013, it paid $14 million to spies who followed me and Pablo. From at least one apartment we are sure we were being bugged. Our website gets so many cyberattacks that we had to hire our own IT person to keep the correct information online. Chevron has paid 2000 professionals around $2 billion to fight us and their spokesman swears that Chevron will continue to fight the case "until hell freezes over", and then we will fight it out on the ice.

It turns out that hell did freeze over and Chevron found itself fighting on ice. The company is responsible for several hellish landscapes and the next stage in Julio's fight shifted from sweltering jungle to the chilly northern tar sand regions of Canada for a David versus Goliath confrontation. The part of David was played by Julio, Pablo, a few native Ecuadorean and Canadian activist leaders as well as Steven Donziger, another human rights lawyer who had been going toe to toe with the firm in the US since 1993.

Despite the years, Steven is still astonished how ugly the fight has become. "When Chevron began to understand that they would lose on the merits of the case, they began to attack us. They rented six PR companies and rolled out a 'shock and awe' doctrine. This military doctrine is simple: show you are infinitely more powerful than your opponent so that he just stops resisting."

Aside from a massive smear campaign, Chevron also went ballistic on the legal front. The laboratory that investigated the soil samples from Ecuador for pollution was sued in one of over 30 court cases that Chevron opened. Steven was one of the other victims, with the firm demanding an astonishing $60 billion in damages from him. An absurd record figure. After Chevron bribed the key witness for $5m in cash and benefits (the witness later admitted that) and managed to secure a verdict from a corrupted judge (who held undisclosed assets of Chevron), they bullied the New York bar to characterize him as an "immediate threat to the public order" and suspend his law license. By September 2018 Steven filed a petition before the Inter-American Commission on Human Rights (IACHR) claiming that US judicial authorities have failed to protect him from Chevron's vicious retaliation campaign. He explains how Chevron tried to imprison him and froze his bank accounts. Chevron also went after his family.

> For weeks there were three cars, with two spies each, at the exit of my apartment. They followed me wherever I went. They even followed my child and wife to school. Maybe they hoped that my wife would feel intimidated and that she would put me under pressure to let go of the matter. But it just brought us together.

Julio Prieto also stayed strong.

> The case was won by us, but there's still suffering because of the lack of implementation. We will not be able to clean the jungle with a court paper saying that Chevron needs to pay $9.5 billion for a massive clean-up. Some people lose faith in justice, but we are determined to prove that those people are wrong. We have no choice. The *Afectados* do not stop crying, getting sick and dying. It's not the case that the problem disappears when we stop watching.

That is something that all frontline heroes in this book seem to have in common. No matter how hard things get, they keep going.

Keep the gas under the grass

Ecuador is not the only place where Chevron is in conflict with tens of thousands of local residents. We move from Ecuador to Romania, one of at least 30 places where Chevron has stirred up conflict.[73] In this case, it is all about gas. And the heroic actions of one man.

When Alexandru Popescu walked away from a comfortable apartment in Ploieşti in the summer of 2014, his 84-year-old father thought his son had gone crazy. Alexandru, a 46-year-old antiquarian, had been cultivating a reputation for doing strange things. Growing organic vegetables on a plot outside the city was seen as odd by friends and relatives, as was joining anti-mining protests. But for his latest plan, his father had only one word: madness. Alexandru had decided to march to Brussels.

On November 25, 2014, after 3 months of walking, Alexandru ended his journey to Belgium with swollen knees and a few missing toenails. He wore a dirty t-shirt that read "No Against Cyanide and No Against Fracking".

Fracking is a relatively new drilling technique that sucks out gas from underground beds of shale. That gas has to be unlocked first, though, in a smash and grab operation that violently injects a cocktail of 596 chemicals deep into the ground, a number of which are carcinogenic. So violent is the process that it causes earthquakes, damaged buildings and gas to seep into groundwater. Drinking water is now so saturated with gas that many people can actually set fire to the water coming from their kitchen taps. This is a messy process that releases a lot of methane, a gas far more harmful to the climate than CO_2. So much is leaking out from US fracking hotspots that a methane cloud is now easily detectable from space. Fracking also requires a huge amount of our old friend sand.[74] In short, fracking is both a weapon of mass destruction and a booming industry.

For me, Josh Fox's documentary *Gasland* at once touched a nerve.[75] Put simply, it made me angry to see what seemed to be the poorer section of American society opening their tap, holding a lighter to the water coming out and a gas flare right in their kitchens. I accept that anger is often my first step towards action. Love can also spur action, whether it be for fellow human beings or the natural beauty at stake. But in the face of the multitude of attacks on the ecosystems we all depend on, anger is more important. In my previous book Walking with Flora, I wrote a lot about bonding with nature and how important it seems to cultivate this connection from a very early age. But if there's no anger involved, what will our resistance to the enemy that attacks us look like? We are not going to stop the Chevrons or Eldorado Golds of this world with a "free hugs" or by sending them for a spell in the natural beauty they care so little about trashing.

Fracking is a good example of what makes not just me but millions of people angry, all over the world. That should be seen as good news. In Europe, resistance to fracking has spread like an oceanic oil spill. As with the Greek goldmines, this has seen moderate citizens converted into passionate environmental activists. Mothers have blocked the path of trucks advancing through English villages like Balcombe. In France and Bulgaria, resistance has resulted in a national ban on fracking. In other places protests have delivered serious delays and sparked mainstream public debate.

Alexandru had fire in his belly when he began his 2900-kilometer odyssey, and for good reason. I wanted to know what lit that fire in him and arranged to meet him near the European Parliament building in Brussels, the symbolic end point of his long journey. In between the tailored suits of the clientele of the posh sandwich bar, an unshaved muscular man appears. My hand only slowly returns to its normal size after his mighty hand lets mine loose. Alexandru is a black coffee kind of guy. No sugar, no cookies and not much by way of small talk. His smile, however, hints at integrity, or empathy. I have a hard time kicking off the interview. Alexandru seems more a man of deeds than words. But after some time, and with the help of Michaela and Daria, two Romanian expats helping the big man to advance his cause, his story emerges, much more of a story than I had bargained for. "I worked for 3 years in the Romanian ministry of defense, so I am not easily impressed." The chuckle that accompanies his statement suggests I need to do some heavy reading between the lines about the things that go on there.

But what I saw in the village of Pungeşti really hurt me. The people there are opposed to Chevron's exploration into gas, which they would get out of the ground with fracking. Chevron initially tried to create social support, but they did that fraudulently. They started with a fake survey, but they had the bad luck that a few of the local people were well informed. Peaceful resistance started with an occupation camp but then on 2 December 2013 I read on Facebook that the riot police had attacked their camp in the middle of the night. Suddenly it was as if Pungeşti was no longer in the EU. The village was surrounded by 1200 cops and no one was allowed in or out. The police behaved like a private bodyguard for Chevron.

Alexandru then tells me how he got through the police cordon and into the village, but a few days after our conversation he asks me not to tell this detail in my book as he realized it might get people into trouble. The whole event did seem to have a profound impact on him.

I decided to go on a hunger strike. I went to the University of Bucharest and stayed 22 days at its entry. This was in the open air, in winter, with temperatures that were dropping to -20°C. Supporters of my cause joined me for 2 weeks, others for 11 days. When my mother became ill, I stopped. But I immediately started making other big action plans.

At that point in the conversation, I steer the Terminator-come-Gandhi figure towards his punishing journey from Romania to Brussels.

I twisted my ankle and had pain in my knees. I lost some nails. But my pain is not important, only my message is. The people who oppose fracking in Romania, they are the ones that suffer. In Pungeşti, they receive fines and many were detained. Their children are traumatized They could not go to school for a long time, while their parents were unable to feed the cattle. All that because they didn't want to have their water poisoned.

By now I no longer need to ask questions, as his story picks up speed and emotion. "I walked alone, but I always had these people in my heart, so I never felt lonely. It also helped me to know that a British activist, known under his activist name Gayzer Frackman, also started a march to Brussels." Frackman handed a letter signed by 205 British NGOs to David Cameron, a letter calling for a fracking moratorium, before going to the European Parliament. Alexandru also participated in a large meeting with Romanians living in the Netherlands, France, Spain, Italy, Germany and other countries, all of whom are members of a movement against fracking or against the planned cyanide gold mine in Rosia Montana, right on a 2000-year-old village that has requested UNESCO World Heritage Status. Afterwards, a delegation of 20 activists met three members of the European Parliament. The group of the Greens / European Free Alliance made a brief video of the meeting, saying that the trips made by Gayzer Frackman and Alexandru were two examples of the growing movement against shale gas and fracking in Europe. European Parliamentary member Stefan Eck for the European

United Left / Norwegian Green Left Group promised to plead for a European ban on fracking in the European Parliament's Environment Committee. Alexandru:

> We were ignored by the Brussels media. But we know that our opponents saw us. Someone wrote an article about us in a magazine for the gas industry. In Romania there were four articles about my trip and my message for Europe, also in the country's largest newspaper. But let me be clear: I did not quite go to Brussels to solve a Romanian problem. We need a European ban on fracking and the use of cyanide in mining and we are building a movement to get that ban.

Alexandru's protest has been lost on the European Commission. Its president, Jean-Claude Juncker, gives fracking a coded thumbs up every time he points to the war in Ukraine and the need to be less dependent on Putin and Russia's natural gas resources. But even aside from the massive problem for the climate: Europe is not the US. People live much closer together and there are few places where you can still explode the underground and thus poison the water table without affecting entire communities. Europe also lacks the extraction infrastructure. Above all, Europe simply does not have the shale gas reserves that the US has. During a European Parliament debate, I hear all this from a scientist delivering a stark message about fracking's limitations to the auditorium via video message. Unconventional gas supplies in the EU are ridiculously low, he says. Even if you invest tens of billions of Euros on infrastructure to tap it at its full potential and try to ignore the popular protests, you would only ever be able to deliver 1 to 2 percent of the volume needed to supply European gas demand. And so the EU did what it so often does: it exported the problem. In the summer of 2018 Juncker pulled a rabbit out of his hat: in the middle of an escalating trade war with the US, he visited Trump and came back with an agreement to make the flow of liquified fracked gas from the US to the EU easier.

* * *

Meanwhile, ever better organized citizens are ready to put their bodies in front of gas facilities. In August 2018 I joined a group who blocked a crucial gas infrastructure of Europe's largest gas field, in the Netherlands. In this case, Shell and Exxon were the bad guys. Gas-related earthquakes had stirred the local Groningers into action a long time ago, but they were now joined by over 700 activists from all over Europe. Mathias (30) is one of the busloads of participants that came from Belgium. This is not a first for him. He said: "Here I do not feel so powerless or alone anymore, in relation to the climate breakdown. But it's not just about that positive energy, our collective strategy is also working." To keep it that way, Mathias gives trainings. "This is about peaceful collective civil disobedience to tackle the climate problem at source: the exploitation of fossil fuels. There is no room for rioting machos here," he said.

This was clear at the last training session before the blockade, where I walk with the Belgian student Stephanie Colling Woode Williams (27). She discusses with her "buddy" what they are going to do if one of them is pushed to the ground by a police officer. Stephanie radiates positivity, calmness and care, clearly helping the woman she teamed up with. Stephanie's affinity group of ten people was just one of ten such groups that jointly practiced a so-called "finger" exercise where a total of about 100 disobedient citizens break through a cordon of a dozen acting cops. At the training three real police officers were watching, but plenty more awaited the activists on Tuesday. Despite some of them beating peaceful protesters who simply sat on the ground with their hands raised, the blockade succeeded in stopping tankers and trains in a bottleneck facility for a full 50 hours. The problem for police forces and for the governments they work for is that activists learn from each other and are getting better at this.

Gas production in Groningen has already caused more than 1000 earthquakes affecting more than 100,000 people living in now damaged to unsellable homes. Compensations are a farce.

Groninger Jan Dales (58) is someone like that. "My father died of a heart attack that I directly associate with the corruption at the institution that refused to acknowledge the damage to his house." Jan paid 2000 euros to a certified researcher who proved that the house did indeed need considerable stability repairs – only to hit a brick wall of unwillingness. With his lawsuit, he made it to national TV and managed to get Prime Minister Mark Rutte to visit him in Groningen. "My troubled lawsuit suddenly caught a bizarre momentum. The day before Rutte came, the judge decided that my complaint was inadmissible, after which the Prime Minister came to tell me that he could do nothing." Jan had made up his mind: "Elections do not change anything here and the rule of law is rotten, so we're left with direct action." That assessment was widely shared in Groningen.

Hanneke (40) is the spokeswoman for Code Red, the movement that organized this action. She saw the will to action grow rapidly. "Code Red was created in the wake of the more established environmental organizations, in whose business model this kind of direct action to keep fossil fuels in the ground does not fit. People are tired of soft action, they want to take action right now. We make that possible." The Code Red action was the largest ever in the Netherlands.

While the Dutch government is preparing slowly, although still too slow, for the end of gas – new houses are no longer allowed to get a connection to the gas grid – many other European governments are busy expanding gas infrastructure. Belgium's cities and communities use public money to invest in new gas pipelines and LNG terminals, through the company Fluxys. The

UK is continuing with a push for gas, fracked or not. European money is used to subsidize new gas infrastructure across the EU and Juncker is a big buddy of Trump when it comes to gas.

Meanwhile, German researchers from the Rosa Luxemburg Foundation concluded that gas is not a so-called bridge fuel, but a bridge to nowhere. With declining stocks, longer supply chains, increasing violence to get at the gas and the fast-increasing resistance in mind, I would add that gas is a bridge to nowhere that is about to collapse. We better get off it before it does so.

Climate Change? System Change!

The action in Groningen was part of a bigger trend. In 2015, around a thousand people stormed a dirty brown coal mine in Germany to halt extraction for 24 hours. A year later, four thousand again stormed it and this time held the fort for 48 hours. This is all part of a broader "break-free from fossil fuels" movement that has spread like a wildfire across the globe. It has attracted citizens of all stripes who have seen through the greenwash of governments and the fossil fuel industry. Like the multinationals they face, this movement is so vast and global that it escapes crackdown by any single state. Oil, gas and coal companies use more destructive power than before in their fracking, deep-sea drilling and tar sands mining. Meanwhile the counter-movement is getting better at the constructive power of mass direct action.

This book is not focused on climate change – which is of course a major issue but also just one symptom of a deeper problem in our relations with this earth. But as the merchants of doubt and most political leaders are so good at spewing fog I briefly have to make sure we are on the same page when it comes to the facts and the challenge we're facing. The debate worthy of such a name about who is the key culprit for rapid change to our climate is over. It is by far and away us, humans. I'm not even going to discuss this, as the science on that is settled. What remains is the debate about the scale, speed and consequences of the problem. Here I argue that the climate breakdown, and given the historic changes a breakdown is actually a better word than climate change or warming, is still underestimated in most media and public debates. Here's why.

The Intergovernmental Panel on Climate Change (IPCC) measures the scientific consensus on climate change and in its progressive reports, the forecasts are getting worse. The sea and temperature rise faster than predicted. One of the reasons why reality is always ahead of the reports is that there are 5 to 7 years between actual measurements and their appearance in IPCC publications. A second underestimation is because politicians have to approve the reports, which puts the whole scientific consensus around climate change under pressure and self-censure to remain within the realm of the politically correct. A third reason for the consistent underestimation is the existence of an army of salesmen pushing only one product: doubt. These are people, such as bribed scientists, who are paid by polluters to question the media, public and politicians about the scientific consensus. They are doing everything they can to try and maintain a long-settled debate about what is causing all the weird weather we are seeing. Their quack science dose of doubt is inserted into political and TV debates on a paid-by-the-hour-to-lie basis, quite literally.[76] In peer reviewed science, the highest form of quality control, only 1 to 2 percent of all published articles raise any doubt about the human causes of climate change. But the merchants of doubt spread their poison everywhere: in the media, universities and even in some big organizations that claim to be environmental NGOs.[77] It aggressively labels honest scientists as alarmists to such an extent that they have grown overcautious and prone to errors of underestimation, according to a 2013 peer reviewed paper.[78]

This potion of corrosive doubt has worked its magic with cold efficiency. It can be traced in the gap between what is reported by scientists to the IPCC and what that body makes of it all, erring on the side of caution and political acceptability. It is again evident in the gap between what the IPCC puts out and what politicians decide to do at climate summits. It all adds up to impoverished settlements like the Paris Accord that do nothing to really deal with the problem. So while the IPCC forecasts a worst case scenario of a 1-meter sea level rise by 2100, a solid scientific publication explained that the sea level has in the distant past risen by 5 meters in 50 years and that it is likely that this will also happen this century.[79] While most climate scientists have already written off Miami and Bangladesh, most politicians and the public at large have not got to grips with that yet.

Another study showed that the temperature could increase by 5°C in 13 years, an eye-watering level of change in terms of impacts, but one that has happened in the past.[80] Most scientists agree that such a sudden increase would lead to a collapse in global food production. In a widely discussed New York Magazine cover story *The Uninhabitable Earth*, top climate scientists were asked what the worst-case scenario would look like. Some express doubt whether homo sapiens will survive at all.

But maybe even more unsettling than worst-case long-term predictions are concrete measurements from the here and now confirming that warming is happening much faster than expected. According to NASA, measurements dating back to 1880 show that the temperature rise has picked up to its fastest ever rate in the last 3 years. The polar ice is breaking record lows. The North Pole could be ice-free around the year 2050, the IPCC warned in 2001.[81] But by 2012, this was almost a reality and sailing a boat on to the North Pole will be a possibility in just a few years from now.[82] Reality is catching up with even the most dire of IPCC scenarios and revealing the merchants of doubt for the charlatans they are.

The really uncomfortable truth is that the politically feasible today is inadequate to deal with the climate breakdown and that the price of this failure could even be omnicide: the extinction of all humans as a result of human action. There's no political failure bigger than that. The greenhouse genie is already out of the bottle and to get that back in the bottle, no amount of Tesla cars or tofu eaters will do. The more you look at the physical science of climate change, the more you understand the inevitability of a political and economic revolution – or death by a thousand cuts.

The scale of the revolution needed is actually rather well known. The Economist was on the money when it wrote: "Either governments are not serious about climate agreements, or fossil fuel companies are overvalued."[83] In 2016, *The Sky's Limit* study showed how many *new* wells we can still drill to get at fossil fuel reserves that we already know about, provided we hold to the agreement not to warm up by more than 2°C. The result of this study: zero.[84]

That knowledge is so disruptive that financial markets prefer to ignore reality for as long as possible. Shareholders and financial analysts want to see all fossil fuel companies hold a 100 percent reserve replacement ratio, meaning they continue to expand their reserves by continually searching for new resources. If we allow them to keep doing that, things will only end in tears.

You cannot convince The Shells and Chevrons of this world with moral arguments about "unburnable" fossil fuels in their portfolio. Their function is to make a profit, not to care for the climate. But it turns out you can convince the Bill & Melinda Gates Foundation or the vast pension funds of countries like Norway. They seem to understand the risks and are just a little more sensitive to public pressure. To keep in line with climate targets that command us to keep around four-fifths of all reserves in the ground, up to four-fifths of today's stock values will need to be written off. That's about 30 trillion euros

(30,000 billion euros). To be serious about climate change means acknowledging the need for either nationalizing or pushing into oblivion and probably bankruptcy most of the largest companies on earth. Either way, there's no longer an economic-shock-free way out of this mess.

The largest pension fund in the world is Norwegian. Worth a whopping €1000 billion, it holds around 1 percent of all shares in the world and almost 2 percent of all shares in Europe. That is more than any other entity, making the fund the most powerful player at the global casino table. Yet despite its power and prestige, its fund managers soon yielded to some intense arm twisting by an international campaign bent on forcing it to drop its dirtiest coal industry investments. That was bad news for king coal, which saw around €5.5 billion of its value go up in smoke with the stroke of a pen.[85]

Norway marked a turning point for a worldwide divestment movement spearheaded by 350.org. It was aided by the Guardian Media Group, one of the world's leading media houses, which has since 2014 run a series of articles called *Keep it under the ground*. It ran one great piece after another on why fossil energy should remain locked up underground. Is this the kind of neutral and objective reporting we expect from a serious newspaper? Is it not too political? I do not think it is. It is merely bringing a scientific reality to the real world and standing up to the army of merchants of doubt, who have been very successful in convincing plenty of journalists and media houses who do not take the time to read up on the latest scientific findings.

The divestment campaign is inspired by a similar campaign against the apartheid regime of South Africa. A campaign against investments in the country certainly helped deal a death blow to the regime. In the US and later in Europe as well, the fossil-free divestment movement has gone viral in universities, cities and churches. But as my own experience showed, the banks still have a long way to go.

Me vs my banks

Here's a somewhat uncomfortable confession: I save money for my pension. Baited with tax breaks, I did what many people in Belgium do and channeled a couple of hundred Euros each year into a private pension fund. I trusted my favorite bank, Argenta, to do the right thing with my money. Argenta is like the ordinary people's bank. They are friendly and speak my kind of language. They have no hidden costs and their offices were simple rather than opulent. Unlike the big banks, Argenta is not even traded on the stock market. A safe haven in turbulent times. I trusted them. I even kind of liked them, for their simplicity and transparency. Then I read their pension fund's annual report. That was the moment I made the horrifying discovery that through that fund, I was betting my money on...Chevron.

In all my naivety, I somehow assumed the fund had been busy investing in things we will sure need in the future: storm protections, old age homes and wind turbines. But in the autumn of 2011, it decided to invest my money in oil: Chevron, Statoil and Premier Oil, an oil exploration company looking for new reserves. I was angry enough to start my own little crusade, dammit.

First to feel my wrath was the Argenta ombudsman. I mailed and asked if Argenta knew about Chevron's enormous pending liabilities in Ecuador and the worldwide divest movement and whether the fund had ethical investment guidelines. A faltering answer suggested the ombudsman was a stranger to this kind of questioning. The fund's prospectus says there is no sectoral constraint. Investments in the oil sector had been made to offset risky investment in the financial sector. Frying pan or fire, in other words. On Chevron, Argenta's ombudsman wrote: "At this stage, however, there is no actual indication that the complaint against Chevron is legitimate." This surely qualifies as an "alternative fact" from before that term took flight. Argenta thus claimed that the Supreme Court of Ecuador and its 220,000-page verdict is not legitimate.

Since my discovery, I no longer save for my retirement. I found that all Belgian pension funds have investments in fossil energy and some are in it up to their necks. On the upside: I know people who, after my first media publication on this, have also quit Argenta's fund in disgust. That motivated me to step up a gear. I would rather be the naive and arrogant busybody than the blissfully ignorant financier of a malignant system. What will we do with extra pension money our unethical funds leveraged when, 30 years from now, there is no habitable planet to enjoy? In the fund's 2014 annual report I find Shell listed as an investment. This is a company that wants to drill oil in the Arctic. Millions of people filed a petition against Shell's plans, including thousands and thousands of Belgians. How many of these people own Shell shares without knowing? How many will sell them once they find out?

A few years later, in June 2015, I launched round two of my campaign to purge my bank of its odious investments. They are still my bank as I need to have an account somewhere and after all, fossil fuels are only a relatively small part of its portfolio. They do quite well, compared to other banks. This time I ask why Argenta still invests in companies like Chevron, Seadrill and Royal Dutch Shell. I refer to a report by the Center for International Environmental Law (Ciel), which warned that rating agencies like Standard & Poor's, Moody's and Fitch continue to make massive assessment errors when rating the value of fossil fuel companies, errors so serious they could trigger a new global financial crisis.[86] For good measure, I add arguments by The Economist, the Norwegian pension fund, the IPCC, the world's largest insurer Allianz, and even the Pope. These are not crackpots, back to the landers or fringe pressure groups, but bastions of the establishment that ought to draw a lot of water among bankers. OK, I am not saying my bank should set its financial compass by what the Pope in Rome says. But just

"transition pioneers". As a member of this Generation T (with the T for transition), I get the opportunity to challenge some of the largest companies active in Belgium. Generation T is a project from The Shift, a large network that brings together NGOs and large companies around what they call corporate social responsibility. That is how a handful of Generation T'ers, including myself, end up at the headquarters of KBC, Belgium's largest bank and recently also Europe's most profitable.

At the tour-de-table we are expected to say more than our name and organization. What would we do if we had the KBC group profit at our disposal? I say that I would invest half of it in lawyers who sue states and companies that do not do enough on climate change and use the other half to switch all KBC investments from fossil fuel to renewable energy companies. Name card delivery: check. During a break-out group I interrupt the man speaking about how they promote cycling to work to talk about their lignite coal investments and those in oil companies with a very bad reputation, like Total. I wheel out the arguments I had honed with Argenta. Again I am told that a sustainable investment fund (not a pension fund) is available, the prospectus for which is probably gathering dust on a shelf somewhere, and that people have the choice to take it. They are doing their best to inform their staff that this option exists, he says. Again, as with Argenta, I stress that people still have no choice in shaping their pension fund and that pension savings are used much more by the average Belgian with his few hundreds of euros a year to put aside than investment funds in sustainable companies for those who have 1000s to put aside at once. He said KBC will think about it and get back to me later.

A little later we are invited by The Shift to lunch with a bunch of CEOs from major companies like IKEA and KBC. I kindly ask Thomas Leysen, Chairman of KBC's board of directors, why the bank scores so badly compared to others when it comes to financing climate mayhem. I tell him about the new cooperative

NewB: 50,000 Belgians who are investing together in the creating of what should one day become a cooperative bank in the hope that one day there will be a bank that will offer products that are not as toxic as KBC's. I flash my NewB card from my wallet, give it to him to have a look. "That's not a bank card but a market signal," I say. Johan Thijs, KBC's CEO, joins us. I seem to have made an impression because Thomas challenges his CEO straight away. Thijs says that KBC will soon propose a new climate policy. From a good source, I also learn that Thomas Leysen contributed to a class action lawsuit against the Belgian State for not doing enough on climate change. The same source tells me that his contribution was modest, but he has been engaging on the climate topic for many years and appears sensitive to divestment arguments.

A few months later, KBC invited me to a presentation of their new sustainability plan, adding that it would soon stop financing coal fired power stations. At the event, Thomas Leysen and Johan Thijs are joined by sustainability manager Vic Van de Moortel and KBC's entire corporate social responsibility team. After a presentation it is time for questions. I am given the honor of going first.

Very nice you stop investing in coal plants. Except for in the Czech Republic. But what about your investments in unconventional fossil fuel operations in fracking, tar sands oil and deep-sea drilling as well as exploration for new conventional fossil fuels? The scientific journal Nature says that this is incompatible with climate agreements. And secondly, very nice that you want to better promote your sustainable investment fund that only makes sense when you have thousands of euros available, but why don't people with an average income like me get a choice? Why are the majority of your customers still stuck with a pension fund that is hastening climatological Armageddon? Why not take this golden opportunity to become the first bank in Belgium to offer a sustainable pension savings option to everyone?

Murmurs rise from the audience and I see some uncomfortable smiles from the panel. Thomas seems rather amused, but passes the mic' to his sustainability manager, Vic. He gives a jargon heavy technical explanation about "small-caps" and that there are too few really sustainable companies to invest their billions in. The argument goes that it would be too difficult for customers to understand that they all fall into one of two categories of sustainable investors. One is green people with a considerable amount of savings who only want to invest in pioneering green companies. Then there are the masses, who would already be happy with avoiding the worst companies. This impoverished explanation from Vic, that such a distinction is impossible to make, assumes there is no marketeer smart enough to segment and target these two groups separately.

The problem is in no way a Belgian problem. Most bankers are still living in a bubble. A carbon bubble. The fact that no president in the history of the US has surrounded himself so openly with fossil energy fat cats as Trump has, is according to some a sign that this huge carbon bubble is about to burst, with the industry engaging in a last desperate attempt to postpone the burst.[88]

Once it dawns upon a critical mass of investors that up to four-fifths of the stock value of fossil fuel companies is tied to what scientists call unburnable fuels and what investors call frozen assets, the shit hits the fan. As we all know from recent experiences: that critical mass ain't big either. It requires only one card in the house of cards to fall – say, Chevron – to get into panic mode and a rush to the exit while you still can. When the carbon bubble bursts, KBC, like almost all banks, is going to get hammered big time.

In the summer of 2018 I get a call from someone in KBC's

sustainability team: they just decided to do it: they will put the first pension fund with no fossil fuel companies on the market. It is now available, I read the prospectus and it's real. One and a half years after KBC's sustainability manager claimed that a fossil-free pension fund was impossible, the bank did the so-called impossible. This was communicated just a few hours after Greenpeace successfully managed to pimp KBC's headquarters right on the day of their annual shareholder meeting. Credit for this policy change should sure go to them and other groups and people who lobbied for this.

But this little personal crusade with the banks sure taught me a couple of things. It seems the ombudsman, ethics and sustainability people, CEO and board chairman of a major bank are at least open to their customers asking awkward questions on their part of funding the climate breakdown. You don't need to speak the financial jargon nor be part of a pressure group to challenge those who you trust your money to. You're a client asking and repeating a legitimate question and that puts you in a position of power, now that the race for the most fossil-free bank seems to have opened up. I think the potential for moving banks away from their exposure to fossil fuel companies is huge and the moment is ripe.[89] So after reading this book, if the urge has taken hold of you, sit down for a moment, read that prospectus of your pension fund and start with "Dear Banker,..."

Profession: Lawyer. Client: Our climate

In December 2014, a group of famous Belgian citizens wrote to the four ministers responsible for addressing climate that the country of surrealism has.[90] Their letter asked the ministers to say for the record that, by 2020, the country would reduce its greenhouse gas emissions by 40 percent compared to 1990 levels, as per the scientific advice. They did not. The episode sparked a class action case in April 2015 against the Belgian state. After some campaigns, the number of plaintiffs increased to almost

40,000. The government raised procedural issues leading to a long delay of the actual court case but also to three victories at three judicial levels just to get the actual case opened. When the lawsuit finally begins, thousands will attend the court sessions. It's going to be massive.

Nic Balthazar, one of the 11 public personalities who started the case, told me: "I think we should not waste too much time on discussing about the hard or soft strategy. Or about the sympathetic or sabotage solutions to the climate crisis. We have to do it all, and the legal side was missing." The importance goes beyond winning a court case as such. Cabinets are already freaking out now that they are being sued and as Nic also said: "A lawsuit is a lot more juicy for journalists than a report by the IPCC. That media attention really is important: it creates the necessary social support."

* * *

The Belgians were inspired by a very similar climate case in the Netherlands – where it all began with one man: Roger Cox. I met him in the European Parliament at an ecocide conference and stayed in touch to get his story on record. During one month, in 20 cities in the Netherlands, everybody who wanted to watch Al Gore's *An Inconvenient Truth* could go and watch the movie for free. Roger paid all the bills. But he will always be remembered as the first person in the world who accused a government of having a climate policy incompatible with international law, or more specifically: as the first who won a climate court case against a state. In June 2015, a judge in the Netherlands accepted his argument and ordered the Dutch State to implement a policy in line with what the science says is needed. A week before that milestone, I spoke with Roger."The cradle to cradle principle shows we can do something about the light bulbs in our office, but that's just not enough to deal with climate change." Roger's law firm did have some money, so they set up the Planetary Prosperity Foundation.

> I began to read IPCC reports, Jeffrey Sachs, Thomas Friedman, Naomi Klein, Jeremy Rifkin, Fareed Zakaria...Then it became clear to me that the influence of the big multinationals is huge. Look at the top 20 of the world's largest companies; all of them are in the fossil fuel sector. That calls for a very strong counterforce. Enter the judiciary.

After writing Revolution Justified, NGOs jumped on it and sent the book to all ministers and the queen of the Netherlands – many of whom reacted. In his book, Roger explained how the state could be sued for not doing what it needs to do on climate change. Soon afterwards, he started the case, with the support of 886 citizens.

Asked about how he juggles his work and private lives, Roger smiles before revealing that his wife had encouraged him to write the book so she couldn't really complain. He also talks about their children as a motivation to do something. But Roger stresses that the lawsuit is not personal and is not about getting his way on policy either. "We only ask the government to have a certain level of ambition, not how it should achieve that ambition. I have my ideas about renewable energy or nuclear energy, but they do not matter. How politicians pursue the ambition that science considers necessary, that's their concern, not mine."

Anger is the first emotion Roger can think of when I ask about his deeper drive. Anger about disinformation. "As a consequence of disinformation, many people still do not understand the effects of the warming. That is why we have promoted Al Gore's film. And if you read how a relatively small group of companies and people steer our ship the wrong way, that too makes me angry."

A pattern emerges as I consider what spurs frontline earth

defenders into action. A love of their children is certainly a factor, but is not what tips them into action. It is more often a certain anger, pent up anger in need of a lightning rod to release its energy. There are no end of causes out there to drive such anger: disinformation about radioactivity; corruption and international blackmail in Greece; Chevron's grotesque shock and awe tactics; the forked tongued merchants of doubt, and so on. The art of being a conscious and caring human in the modern age, it seems, lies in taking anger that threatens paralysis, depression or madness and finding a cause that converts it into a force for good. Maybe anger should not automatically be seen as a negative, but rather recognized as an essential energy source driving frontline heroes on.

When well directed, anger is followed by joy. One week after talking with Roger, I follow the court verdict in the Netherlands through livestream. The tears of happiness, the hugs and the citizens beaming with joy flood my living room in Belgium. The judge simply ordered the Netherlands to reduce its emissions more than it had or was planning to do. The state appealed, but in October 2018 the verdict was confirmed. Roger's victories activated counterparts in Belgium, Australia, Brazil, Austria, England, Ireland, US and Norway, all of whom now have the jurisprudence they were waiting for.

* * *

I talked with Roger Cox and Nic Balthazar not just about the court cases themselves, but also about their mental fights. There's no point denying that those who dig themselves into the climate change issue are faced with not just the limits of the planetary system but with the limits of the human brain itself. Psychologists like Harald Welzer analyzed how our brain tries to keep these facts out that are just too unsettling – whether it is an ongoing holocaust or an ongoing climate breakdown. So how do they deal with that? Nic Balthazar told me something I've since used quite often in conversations on this topic: "The best way to get me out of that depression is to do something. A doctor who sees a cancer will not start crying, will he? He will wonder what he can do."

Chapter 3

Above ground: Life on earth

The white man's burden

Imagine a young man with an odd skin color suddenly standing in your vegetable garden. In his right hand is a device full of buttons and a screen. In his left hand, a white paper filled with boxes in all shades of red. Imagine this intruder looks to his device, to his paper, to your vegetable garden and back to his device. He then takes a pencil from behind his ear, scratches something on the paper of his and pushes a button or two. Then he walks on 30 meters and repeats his ritual.

That man was me, in 2002, in some godforsaken volcanic part of Uganda. The Ugandan at the edge of the Ntambi crater lake who witnessed this scene approached me and asked what the hell I was doing in his vegetable garden. At least, that is my best guess at what he was saying in Swahili or whatever language he spoke. I was simply doing research for a Master's thesis on land use changes that involved comparing different satellite images.

Look, you see that red block on this page? That corresponds to the banana plantation over there. This device tells me that I'm on this red block on my paper. This sheet is the infrared radiation of this area. That rosy block over there will be your meadow there. Less biomass means less infrared radiation and thus rather a pinkish color. That's how I map this area. Cool, don't you think?

I probably understand as much from his reaction as he from my story. But I take the furrowed brow and disgruntled sounds as signs of trouble. In hindsight, who would not be pissed off if, all of a sudden, a black man in weird garb is found wandering through your garden taking strange measurements. I decide to hop it and he decides not to chase me off his land. I was probably taken for *mzungu*, a slightly crazy, but otherwise harmless white person.

There were other uncomfortable encounters during my fieldwork in rural Uganda, but this one really left a knot in my stomach. Was this expedition more than just a childhood dream of mapping remote corners of the world coming true? Why was he so angry? Was my dream his nightmare and if so: why? A peasant farmer somehow pulled me from my cloud to his situation of abject poverty – which would remain not only unsolved but even unseen, after we enriched ourselves with the data we needed, in his garden. He confronted me with simmering doubts. What did he make of me and our team of three white men and a white woman, trekking across his land with canoes and mud sample tubes, heading directly for the holy crater lakes? What good could ever come from four *mzungus* boring holes in the realm of the gods?

When I read Frank Westerman's book Choke Valley many years later, I got closer to answering these questions. We as scientists were arrogant trespassers, meddling with lakes that are the pantheon of their gods. After reading Choke Valley I'm pretty sure that if something bad happened to the village during our presence, the wrath of these gods would have been a powerful narrative and put us in serious trouble.

Dirk, the expedition leader, never seemed to descend from the cloud called scientific research. The mud samples we took from the bottom of the crater lakes would allow us to reconstruct the local climate stretching back hundreds of years and learn the cycles of droughts and floods. He justified our invasion for data by arguing that our scientific foray would eventually benefit local farmers. I found it easier to see a direct connection to our career prospects. After all, none of our team would later explain to the local people how to use our information in their favor but all of us would go home advancing our careers thanks to these data.

Dirk later asked me to do a PhD on land use change around crater lakes in Central Africa, to help his research. As a 22-year-old student, the offer was almost impossible to resist. I loved the adventure of these expeditions. Adrenaline would sometimes keep me up the whole night. I was finally mapping pieces of earth, something I had dreamed of doing since childhood days spent in my grandfather's treasure room, where he kept a cabinet filled with his maps and a mammoth Times Atlas. I can still recall the feeling from when I was pouring over them. From the comfort of his desk, I began making my first mental trips to the edges of the world. I crossed the Siberian rivers Ob and Lena looking for the least-mapped regions that were in need of a cartographer. From a very early age I felt the urge to go right into the great unknown, to beat all the odds and come home with new knowledge about this wonderful planet of ours. As a teenager I classified thousands of stamps from exotic corners of the world, kept all foreign currencies I could take hold of and never missed an opportunity to score a national geographic magazine from no matter what year. In hindsight, I had prepared for this moment for all my life. It really was a childhood dream coming true.

But the more excited I got, the more I also wondered about the fate of the Ugandan people we met. They had zero interest in my fantasies. Why should they? It seemed so selfish. It was selfish. When I returned, I decided not to write a doctorate, but to study something different: development aid. And with that, the only disease I caught while in Uganda was an infection that does not feature in the annals of tropical medicine; it was the white man's burden.

The white man's burden was the title of a poem by Rudyard Kipling, a work that served as a call to civilize the "less-developed people" of Africa, in particular. Today, it is known as an ideological underpinning for wholesale racism, colonialism and imperialism. It was up to the superior whites to lift the darker races out of the swamp. The mindset persisted even as quasi-religious colonization morphed into pseudo-science development programs in the mid-twentieth century. A patronizing mindset was sanitized by a thin layer of philanthropy and altruism. From now on, the white man's burden would be morphed into a strictly noble task. Wise whites would guide ignorant blacks to advance their civilization – with the assistance of loans at friendly rates. But the motive driving certain private companies has been less high-minded. They have sucked the blood of this romantic paradigm for all it is worth. They count on white man's burden patients like I became when I was 22 to do the fieldwork in Africa for them. They need employees who believe in this myth. But what they have created is a global unequal exchange of goods, whereby Africa has become the cheap provider of raw materials for others. More worrying still: this is a trend that only grew once decolonization came to an end. That should set off alarm bells about the hidden agenda behind so many projects in Africa.

The white man's burden is a piece of the puzzle to understand why we whites succeeded in allocating much more than our fair share of natural riches to the West. While I wasn't exactly doing the prospection for mining companies or land grabbers, I was doing a rather similar mapping in Africa job and was busy building up a good CV for just that kind of work. It was many years later when I started to connect the white man's burden, the urge to map the unknown, capitalist expansion and a wave of terrible land grabs going on in Africa.

Lying labels

What was going on around the Ntambi crater lake in Western Uganda in 2002 was in a way very similar to what happened in much of Sub-Saharan Africa by 2008. A global food crisis was reaching a climax. Prices of staples such as rice and wheat had gone through the roof. Foreign actors were buying vast areas of farmland. Measurements differ, because many land deals are not transparent, but Professor Saskia Sassen put the figure at more than 200 million hectares of land bought by foreign firms and governments between 2006 and 2014, with a peak in 2008.[91] That is an area equivalent to almost 70 Belgiums or eight times the United Kingdom. And there are still many deals in the pipeline. Big names from the Brazilian agro industry had wanted a 35-million-hectare stake in Mozambique to create a farm ten times the size of Belgium.[92] On their satellite images, Mozambique had seemed the ideal spot for a giant soy farm. Soy for export, not home consumption. Soil scientists and other technical experts gave a thumbs up. They probably prospected the land just like I did in Uganda: connecting satellite imagery with the local situation – minus the population. At that point, obstacles on the ground, such as 100,000 local residents, were merely problems to be fixed. You promise them work. They will only find out that it is slave work when it is too late anyway.

In practice, land grabs impoverish locals to enrich shareholders on the other side of the world. The precedents are so numerous that this can be stated with confidence. Let us take an example that is well known. In London, at the headquarters of the New Forests Company, someone decided to transform "abandoned land" in Mubende, Uganda into a cash cow. In England or Europe, you cannot earn money by growing forest on farmland. But in Uganda, this makes you profitable and a role model. The New Forests Company became Uganda's investor of the year. Its Ugandan plantation complied with Forest Stewardship Certificate (FSC) standards. Less than 10 percent of all the world's managed forests meet FSC standards, the gold standard for ethical timber production. The project should have been a safe bet.

The reality was different. More than 22,500 people were evicted from land they had been farming for generations, according to testimonies. Oxfam recorded the horror associated with the construction of this "exemplary" tree plantation.[93] To ensure nobody tried to return to their land, soldiers destroyed the school, set fire to several houses and in the process killed an 8-year-old.[94] Families were forced to leave on foot, carrying whatever they could wear and without compensation.

This horrible episode caught the attention of big media like The Guardian, The Wall Street Journal and Al Jazeera.[95] The World Bank ombudsman stepped in and negotiated a compromise, 2 years later. The New Forests Company gave money to a cooperative to help the affected people to new land and new infrastructure like schools. But the community will never regain the living standards it has lost. If this is a promising practice by the investor of the year and later a successful campaign to prove that the FSC logo is working, then how bad is it for those affected by the thousands of other projects that failed to gain media attention?

Beatings, scorched earth tactics and eviction of communities without compensation. The Ugandan project is a win-win of sorts, except the London company wins twice: once for the FSC wood and once for the carbon. Trapping the latter through tree planting earns carbon credits, credits that have been invented to keep our lifestyles unchanged while the difficult work of reducing emissions is left to the poor. Also invented to keep up the pretense that we are "dealing with the climate problem". This is how the so-called solution to a global problem created by industrialized countries poses a new problem for the people in non-industrialized countries, who are often already most affected by the first problem.

The main argument of the development community is that the project provides jobs. Indeed, the brutally deprived people were offered work. Not for the promised $100 per month, but for $30. That is one dollar a day. "But they did not even have that dollar beforehand," is what the development crowd will say. In the past, these people were self-employed. Farmers rich in self-esteem, togetherness and everything that is essential to them: crops, animal products, plant medicine, fresh water. Now that they are "developed", they exchange all that for a dollar a day. That is good for the World Bank statistics: they will show that thanks to "development" people have gone "out of poverty".

Needless to say, this story sounds different at the consulting company that gave New Forests Company the FSC label. In this case the consultancy happened to be Société Générale de Surveillance (SGS), which was paid by the New Forests Company to give it the label. Why should SGS, a company that depends on profits, bite into the hand that feeds it? Of course, to maintain the illusion that there is also the certifier of the certifier. Accreditation Services International (ASI) checks the certification companies. In Uganda, they also checked SGS and concluded that all was by the book.[96] But if scorched earth tactics, torture and expulsion are in the script, why do Western consumers pay an additional fee for their FSC wood?

A tougher question is whether it would have made a difference if ASI concluded that big mistakes were made? The watchdog organization FSC-Watch claims it would not. ASI has identified multiple violations that have all gone without penalty, it says. Phony certificates continue to be awarded.[97]

I check the FSC story with Filip De Bodt from Climaxi, the Flemish chapter of Friends of the Earth. We have known each other for years and if there is one activist in Belgium with a specialization in exposing greenwash, I think it's Filip. "As a label agency you have to choose," he says.

Either you are strict and expensive, or your criteria are flexible to make sure that the consumer buys "green" for only a small additional price. Most labels start with good intentions but then make the latter consideration. Most of them don't even count the number of kilometers their product is traveling.

FSC was also founded by NGOs with good intentions. But when the label moved from paper only to wood, it began to allow industrial tree plantations to apply for the logo. These are anything but sustainable.

Filip explains who is behind FSC labeling:

The WWF is the driving force behind most labels such as FSC for wood and paper, but also for fish, soy and palm oil. And who is on the WWF board? People from banks and industry, such as Coca-Cola. As I see it, WWF is the environmental organization of the employers and an instrument for large companies. Their hidden agenda is to maintain the flow of biomass to the core consuming areas of the world. Securing the supply chain is about securing profits and they feel that our increase in consumption should not be jeopardized. Now that there are more critical consumers, more sustainable packaging and propaganda is needed. The labels are necessary to keep the reality of an ever-larger attack on our remaining natural capital hidden. The labels also function to kill debates. The biggest enemies of label-producing green NGOs such as WWF, as well as Conservation International, are the people who question the whole chain and who want to replace labels with a peer-to-peer model of lending instead of buying, a local model or a circular economy.

Resistance to labeling often brings awkward tensions, according to Filip.

Unfortunately, few people ask the open question whether those labels do not contribute to maintaining a flow that inherently can't be sustained. Within the NGO community, questioning labels is like cursing in church. When we published our report Fish and Run, with a critique of the WWF promoted label for fish, this caused a lot of tension and attempts to cope with critical sounds within the environmental movement through a so-called *pax ecologica* – to silence us. At Climaxi we look beyond the labels. For example, we set up a solidarity campaign to import a large amount of Greek products that come from a good cooperative. They produce in a fair way, but don't have the money to label their products as such. I see a future in bringing together production cooperatives like them across Europe. This already happens in the fisheries sector, with pro-life.

It also exists in the renewable energy sector with REScoop. Filip: "The free market is problematic but nationalizing everything is not the solution. I believe more in connecting local cooperatives and control from below than in bureaucratic socialism."

Filip is not alone in criticizing FSC. Anyone with a little understanding of the FSC field and free from conflicts of interest says that they want to dismantle the FSC myth. Those in NGOs that established or maintain the FSC system defend its label. They have to do that, of course, out of loyalty. An FSC boss gave a scathing critique of the system he presided over for years, but only after his departure. He said that the entire FSC certification system is a myth.[98]

The report *Who watches the watchmen?* did for the RSPO label (Roundtable on Sustainable Palm Oil) what Filip and others did for the FSC label: destroy the myth.[99] None of the big NGOs dare to admit the mistake they made and keep making. Their story goes like this. The label encourages good practices, is already better than all the rest and if there is a problem, we will solve it.

Simon Counsell of the NGO FSC-Watch does not buy the bad apple narrative. "FSC members have lost control of their own label." he says. "The real power lies with the certifiers and they are private companies and only interested in selling as many certificates as possible. I do not know one FSC label that was withdrawn as a result of a formal complaint." This quote comes from a strong piece of research journalism called Sustainable on Paper, written for one of Belgium's top magazines. It sure caused some long overdue debate in Belgium, with many media jumping on it.[100] One of the authors, journalist Leo Broers, recalls the period with interest.

> My experience with most media is that our story and the criticism of the FSC was usually well received. But our content was often twisted and the articles in the press always ended in favor of the FSC. Journalists simply have a lack of time to do some fact checking and they also suffer from a bias to end articles with a positive note. The story usually ended with the FSC apparently learning from its mistakes. Of course, the FSC used its right of reply to spin the case and to dispute facts. But the regrettable truth is that nothing has changed in the field and that the label must be maintained to mask our overconsumption and sweep our environmental crimes and human rights violations under the carpet.

Leo and co-author An-Katrien Lecluyse were told to report their findings through the official complaints procedure. That is exactly what they did, only to find 2 years later that the complaint procedure is also biased in favor of FSC. In a follow-up article, the authors exposed how a firm wiped out 1200 hectares of Atlantic rainforest, yet was still awarded a so-called ethical timber label from the FSC. "The language used in the certification reports is also interesting," Leo said. And here he gets to the fundamental issue with labels like the FSC.

> Certifiers talk about "illegal occupiers", while those people have land titles. In Africa, like in South America, it often happens that the same piece of land is claimed by two or three owners. And the certifiers always choose the side of the

company that requests the FSC label. That is the core of the problem. And that is confusing for the consumers in the West, who believe that FSC is in favor of the rights of local residents and for nature. In practice, the FSC label not only violates their rights, it also kills the debate about it.

The very *raison d'être* of the FSC – and similar labels like RSPO for palm oil – is that the consumer pays a premium to *not* have to bother with debates about the injustices in the supply chain. Take away those consumers who don't care plus those who've already paid to prove they do care and few people are left to debate with.

One of the founding organizations behind FSC, RSPO and other labels is the World Wildlife Fund or WWF. According to them, the good news is that RSPO palm oil was produced in an environmentally friendly way, "so we should not give up on those products".[101] To "not give up" seems to be at the heart of the creation of WWF. This organization was established by rich European princes, who feared losing their hunting rights as countries in Africa gained independence. So it was no coincidence that the honorary president of WWF Spain is the country's king. He's holding the traditional WWF values high. That is, until he fell and broke a leg while on an elephant hunt in Botswana.[102] Another line of founding fathers of WWF are the Rockefeller family, who made their fortune with oil. With such origins, how did we come to see WWF as *the* organization protecting life on earth? Even their President and CEO comes from Mars.

* * *

Palm oil is only cheap thanks to the cheap evictions of animals and humans, extortionately low wages, cut price transportation (more on that later) and the willful neglect of the environmental costs of its production. As it is only on the back of these injustices that palm oil has boomed, I believe we do need to talk about giving up a practice that we developed and later justified with some window-dressing labels.

Palm oil is also blended with fossil fuels to make biodiesel, along with rapeseed and soy. While Western governments continue to back biodiesel as sustainable, they are making insufficient investment in real renewable energy, according to Anne van Schaik of Friends of the Earth Europe. The EU's policy of using biodiesel for transport has actually increased its overall transport emissions instead of cutting them, according to a highly embarrassing study from the European Commission itself.[103] It found that because virgin forests are churned up to grow crops for biodiesel, that biodiesel actually emits more greenhouse gases than ordinary diesel. Because most people associate the term "bio" with "good for nature" the word "biofuels" is consumer fraud. Biofuels are an "alternative fact" to borrow a term from the Donald Trump establishment, used in EU propaganda to show that it is busy "saving the climate". EU countries pump billions of tax money into what is not just a fake solution, but a net transfer of biomass from the tropics to car users in the EU. When it comes to biofuels, EU countries act as vegetarian bloodsuckers telling its flock that sucking the lifeblood from others is an act of sympathy and sacrifice.

It is not just our tongues, hair and cars that are drenched in tropical blood. An increasing amount of our stoves deforest tropical jungles and displace people and animals as well. Having a warm house is a basic human need, but one may wonder if it is a good idea to chop a rich primeval forest in the Amazon, plant an industrial tree plantation and make wood pellets out of that to burn in the EU. Fuel from pellets as a residual product of a sawmill seemed like a good idea at first. That biomass would otherwise be "lost", even though there are many other applications than heating. But today, the demand for pellets is so high and European subsidies are so generous that pellets are now being imported from Brazil to Europe, some carrying the FSC label.

The next step for the marketeers of green capitalism is to reassure consumers. Do not worry, they will say. The chainsaws cutting up that ancient rainforest for your pellets were running on ethical biofuel. And all of our products are panda-free.

* * *

Personally, I found one figure pretty shocking when it comes to the problem of tropical biomass extraction for use in industrialized countries: industrial tree plantations have grown fourfold, from 15 to 60 million hectares, in just 20 years. This has mostly been at the expense of rainforests and mostly for products exported to industrialized countries – such as paper. In a tropical tree plantation, whether it is for paper, palm oil, biofuel or pellets, biodiversity is reduced from staggering diversity down to close to zero. People who used to harvest fruits and medicinal plants and hunt in the forest lose everything they need for survival. We get the things from the shop, pharmacy or simply by turning on a tap. But for them, it is, or was, the forest that provided it all.

The fact that the Food and Agriculture Organization of the United Nations continues to classify plantations as forest is a scandal. Hundreds of organizations and scientists from all over the world wrote an angry letter about it.[104] A tree plantation is a green desert, not a forest. If a plantation is a forest, then North Korea is a thriving democracy.

Let us take a step backwards. Demand for wood and derived products has exploded. Labels facilitated that expansion. The EU supports the rapid growth with mandatory targets and its member states put billions of subsidies into expanding the use of wood from other continents. In 2012 alone, the EU also imported €6 billion of meat, leather, palm oil and soya produced on land that used to be occupied by tropical rainforest.[105] Between

1990 and 2008, the EU was the world champion of importing deforestation products, far ahead of North America and China.[106] The money spent on policing the +100 billion dollar a year market in illegal trade of wood is a bad joke. All of this makes it fair, in my view, to call the EU's policy regarding the world's trees a subsidized ecocide, repackaged and resold to the general public as an altruistic sacrifice for the greater good.

Water, the source of life

The story of humanity on planet earth is closely linked to who gets access to what parts of fertile land, biomass and water. Water shortages will become one of the major twenty-first century crises.[107] In some areas in Spain, the crisis is very tangible.

For a man who came to Brussels to get attention for the ongoing ecocide in his village, 66-year-old David Dene leaves a happy impression. This British globetrotter radiates energy and happiness. Since the late 1990s, David has lived in Rio de Aguas, an eco-village in the hinterland of Spain's Almeria region. He may look like a hippie but the highly educated David convinced the European Parliament, European Commission, the EU Petitions Committee and the UN's Human Rights Council of his case. I met him during a debate in the European Parliament on getting ecocide recognized as a crime against humanity and was intrigued by the man.

The Rio de Aguas Valley has been inhabited since the Neolithic. Just 7 kilometers away, westerns are filmed in the barren Tabernas desert. But Rio de Aguas is blessed with an abundance of flora and fauna thanks in large part to numerous groundwater springs. Compared to the desert next door, it is a little garden of Eden. And it's not the only oasis in the (semi-) desert. Some 35,000 people depend on the aquifer that David is defending with everything he's got.

Rio de Aguas saw its population rise in the late-twentieth century. This was mainly due to an influx of European immigrants who started a new eco-friendly life. A life with due respect for the fragility of the local environment and with a few new technologies such as solar panels. A few dozen people turned one ghost village into an international meeting place where knowledge was shared on how to build an eco-village from scratch. Rio de Aguas was flourishing once more.

That was until the Ministry of Environment declared the country upstream suitable for agriculture. A license for an industrial olive tree plantation on 3600 hectares was issued. Businessman Juan Carrion Caceres bought 2300 hectares and planted millions of olive trees.[108] The over-exploitation culminated in 2017 with an estimated 6 million super-intensively grown olive trees, who drink 10 liters of water per day and per tree. That's 60 million liters per day going to the olives.

Now, there is a lot to like about olives. Mediterranean people are reputed to live longer than average thanks to a diet that is big on olives. Olive trees are also a cultural heritage. But planting 6 million thirsty trees in a semi-desert with a very delicate water balance cannot go without consequences.

The new trees drink around five times more than the total rainfall across the entire valley. The only way for this plan to succeed was to quickly use the groundwater reserves that had accumulated over millennia. The business model boils down to water robbery. Groundwater and surface flow levels in Rio de Aguas have declined rapidly since the planting of the first half million trees. The ecosystem began to resemble the nearby Tabernas desert. "Bulldoze, poison, pierce, suck water and harvest the crop. That's the summary of the business model," David says, visibly upset. The over-exploitation of the aquifer is already causing a new exodus, with populations of nearby villages collapsing quickly.

To my surprise, David quickly regains his calm. "I don't believe in bad people and bad intentions, I don't condemn anyone. I can only assume that some people are not yet aware of what is happening".

The story of Rio de Aguas fits a pattern that forest communities in Brazil, Cameroon or Indonesia will recognize. As usual, it is an unequal battle. David Dene:

> In Andalusia, contracts are often given to friends and families, to companies closely linked to politicians. Threats from them to not complain are normal here. As a result, most people don't want to be mentioned in any form of protest, not even in a petition They are afraid to fight the rich and powerful. But that's exactly why you have to make it clear to this mafia that you are not alone in your protest and that there are thousands of people behind you.

David reminds me of the biblical story of young David in his battle against the mighty Goliath. In that story too, David was not alone. A large army was standing right behind him.

But how do you get a campaign rolling from a few pioneers to a mass movement?

> We had some rain in 2012, so the situation with the water was a bit better. But 6 months later the river ran empty, as if someone had pulled the bath plug. I emailed pictures to Almeria University and in July 2014, Professor Jose Maria Calaforra confirmed what the villagers had long known. The new olive plantations are extremely harmful to Rio de Aguas.[109] But as early as 2011, a report funded by the EU showed that our aquifer is overexploited by 330 percent.[110] I started a campaign and 3 months later, the Facebook page, Ecocide in Rio de Aguas, counted 23,000 followers. We distributed 7000 leaflets, got local newspapers interested and received support from a global End Ecocide coalition of NGOs and experts. Legal support arrived from various countries. We have experienced that if you make a clear call

for help worldwide, many people answer that call.

One organization that joined the fight was The Grupo Ecologista Mediterráneo, GEM or Group of Greens in the Mediterranean Region. They opened a criminal investigation into one specific part of the olive tree estate, which was planted without an environmental impact assessment. You don't need to dig into the legalistic arguments to understand that this estate is more like an industrial activity than a traditional agricultural activity, and that the impact of this industrial activity on the water balance can and should be measured.

But how does David deal with all the destruction of this little paradise on earth he helped to create, on a personal level?

"It helps that I worked in Ecuador for a long time, where the situation is much worse. The leaders of the resistance there are incredibly brave and strong." How true these words are became clear soon after our interview. José Isidro Tendetza Antun, one of the Ecuadorian friends of David and a leader of the Shuar, a group that opposed the Mirador mining project, was kidnapped on his way to the 2014 climate conference in Lima and tortured to death. David sees him as a martyr for the Ecuadorian Amazon, and for the Shuar population.

It's great to see how the Shuar live in harmony with mother nature. They treat Pacha Mama like their mother. I do not see how I could feel depressed when I see how those people in Ecuador live, so respectful while under so much pressure. For me in my fight in Spain, it also helps that there is a strong sense of solidarity here in Rio de Aguas. I live in an international circle of friendship, trust and help, a circle that keeps fear at bay. Our whole campaign is about facts and awareness. It's important to enjoy what we do. We are planning a music festival for 300 to 500 people because our resistance is like a party. We are going to enjoy the intensity of this experience of cooperation, at every step of the way.

India's water millionaires

The state of the water table is often a good measure of the state of prosperity. This is the case in Rio de Aguas, but also in most of India. The small village of Hiware Bazar realized this very well, changed how it did things and as a result, its average income is now 20 times higher than 15 years ago. This is an example that shows what kind of prosperity is possible if we live in harmony with nature.

Hiware Bazar is located in the Indian state of Maharashtra. In 1995, more than 90 percent of all residents lived below the poverty line of less than one dollar a day and it was dependent on federal development programs. Twenty years later, it has no poor people. Their recipe for success has three ingredients: investing heavily in the local ecosystem, building a strong local and inclusive democracy, and focusing on a long-term vision. The three are inextricably linked. In a country with 4 very wet and 8 very dry months, you mainly need to find a way to keep the soil fertile and moist for as long as possible, fight erosion and floods and prevent water shortages. A series of very small earthen dams and reservoirs made by the villagers catch the water during the monsoon and avoid soil erosion during downpours. But the village council of Hiware Bazar did more: it decided to intervene wherever it noticed a threat to the water cycle. The village issued a ban on cutting trees, an obligation to replace channel irrigation with drip irrigation, a ban on water-intensive crops and, finally, the replacement of goats by cows. Goats pull up plants including their roots, causing more runoff erosion of soil that goes on to fill the reservoirs and threaten the whole system. The village argued that to succeed, you have to be holistic. So after intensive discussion they decided to invest their village's 5-year federal development grant not in roads or schools but entirely on water management. This stimulated an immediate employment boom, with landless families given priority access to paid work. The decline of the water table was reversed and, as a result, the grasses flourished and fodder yields for livestock soared by a factor of 60 between 2000 and 2004.[111]

Agriculture and animal husbandry took such a positive turn that migration to the overcrowded Indian cities ceased. Now, each year the village council makes a water budget, one that cannot be written in red ink. It is clear to them that their wealth is gauged less by money, and more by the security of their biomass, soil and water resources. Ecological economists call the availability of these resources the *GDP of the poor*.

The people in Hiware Bazar and Rio de Aguas both understand that water is the basis of a good life. They know that environmental care is not a high-minded intellectual fad. They are the living proof that debunks the myth that you must first be rich before you can afford to care about the environment. According to this Western maxim, citizens must first meet their basic needs, whose definition can then be stretched to the extreme. But in Hiware Bazar, Rio de Aguas and so many other places, reality shows the opposite: citizens must first have a healthy ecosystem. In their peer reviewed article *Is India too poor to be green?*, Leah Temper and Professor Joan Martinez Alier take apart the idea that environmental care is only for the rich.[112] They show that the West did not clean up its act. It merely exported the polluting production of its goods to countries like India. Martinez Alier is also the author of the book *Environmentalism of the poor*: a heavily underestimated area in the environmental movement, especially in the richer countries. While the poor usually don't frame their actions as environmental, the health of our planetary ecosystems depends a whole lot more on their successes than on the environmentalism of the mining CEO who switches from a Hummer to a Tesla car.

In 2006, I spent a few months in Indian villages in the state of Chhattisgarh, where I examined micro-credit systems controlled by women's groups. After talking with hundreds of villagers a certain pattern emerged: most of these women were sensible long-term thinkers, loved engaging in cooperation and managed the scarcely available raw materials in the most sustainable way possible. Too often they were just not allowed to implement their ideas – usually by a group of male short-term capitalists with more resources and better political connections. These women did not need anyone to deliver to them "development" from some external source, they needed more of an opportunity to do what they do well. They needed access to resources like water. I recall how a group of women told me about their fight to manage a fishery in some local ponds – involving a stick fight by dozens of angry women against some wanna-be outside businessman that had set his eyes on the same village pond. The women won the stick fight, got rich from selling the fish and the village was flourishing because of their micro-business.

Amartya Sen would say the *unfreedoms* of caste and gender should be broken. A local NGO that brought the women together in small saving groups could help to do that. No patriarch could stop them once a level of confidence had been reached in the group. But however far sighted they may be, the sad reality is that sustainable ecological management in villages around the world is increasingly threatened by global forces and the mega projects that come with it: large dams, mines, factories and plantations. These trends have come to light in recent decades as capitalism has expanded. The West's "green consumption" is not always helping; on the contrary, because in the unconscious mind of too many green consumers the conclusion will be that they have bought off any feelings of guilt or care for the bigger injustice of the system of unequal ecological exchange they still benefit from, at the cost of others. If there is no global democratic correction to this destructive process, can we expect undemocratic forms of reaction to gain more ground? Is it not naive to assume that we can continue to benefit from the wholesale plundering of others without any consequences?

Europe is like an adolescent in their thirties still living comfortably in Hotel Mom and Dad. The parents still make their spoiled kid's food and clean up their rubbish, but are growing quite literally sick and tired of doing so.

What I learned about water in Pakistan

After Hiware Bazar and Rio de Aguas, I am ending this journey along water's frontlines at Borith Lake. This little lake is in Pakistan, a country that lives in a tense standoff with its neighbor, India. With one and a half billion inhabitants in need of water, it has not escaped political notice. When tensions between the two countries rose again in the fall of 2016, India's Prime Minister Modi weaponized water (again) by threatening to divert the one key river flowing from India into Pakistan. After Egypt, no other country in the world is so dependent on one river as Pakistan, a river that has its source in the lands of its arch enemy. The Indus is a lifeline, a miracle and a key origin of our civilization. From space, you can recognize the Indus as a long blue and green snake winding itself through otherwise brown desert. So if Modi says he will divert the Indus, he is brandishing something worse than a nuclear warhead.

In the spring of 2008, I traveled along this Indus, on the Pakistani side of Kashmir. I was on honeymoon and the trip rarely felt dangerous. But on this stretch we grew scared. It had nothing to do with religion, Taliban or bombs. At the time, Pakistani Kashmir was still relatively free from these troubles, but sadly not free from water and climate change troubles.

Above Karimabad the Ultar glacier sends its meltwater through a narrow gorge. The only way through is to follow narrow irrigation channels built into the vertical cliffs on either side of the canyon. This is not the trip for anyone suffering from vertigo. A squat wall channels water along a cresting canyon wall. On the other side is a sheer drop into the abyss. According to the locals, more than 100 people have died while maintaining this channel. Yet it is essential to keeping Karimabad the oasis it has been for centuries.

Scoping out the perilous route ahead of us, we come across a man armed with bags of sand and a shovel. He succeeds in sealing a leak and then rushes to the next. For us, covering this stretch is a personal choice and a bit of an adventure (although more than we asked for). But for the Kashmiri living here, it is a matter of life or death for the community. A few days later and a little further downstream, we see at Borith Lake what happens when an oasis community in a mountain desert is cut off from the glacial meltwaters. A canal and fields lie dry and idle. We learn that nine families had to abandon the area after the canal ran dry. Despite daily efforts to keep the waters flowing, they simply could not catch up with the rate of melting and thus the ever-changing position of the entry point of the channel.[113] The Himalayan mountains are warming three times faster than the world average. When the channel is not fed with water, the people are not fed with food and flee. Yet it had all looked so promising on the billboard next to the lake. A map shows a new village with 125 pieces of land, canals, a school, a medical post, a playground and so on. Pakistan invests a lot of money in Kashmir, but not enough to reckon with climate change. This cruel fact would become even more apparent when the Indus flooded in 2010. This was a flood beyond our normal understanding of the word. A total of 20 million people were affected – almost twice the population of Belgium, $44 billion of damage was caused in what is a poor country. And of course much of that year's harvest was destroyed. Climate scientists later linked this epic disaster to climate change.[114]

The odd silver lining of this mega flood was that the dying Indus Delta finally received some water again. In recent decades, it has dried out as all the water of the Indus was used before the mighty river reached the sea. In their period as colonial masters of the Indian subcontinent, the British laid 100,000 kilometers of irrigation channels around the country, much of it around the Indus. The waters nurtured cereals, as well as cotton for the British textile industry. Britain imported significant amounts of Indian cotton between 1815 and 1900.[115] Pakistan inherited this vast infrastructure, but also a fast-growing population to feed. By using so much water for cotton and food production, the Indus Delta dried up, which drove hundreds of thousands out of the delta and into urban slums. In the 1990s, Karachi became the fastest growing city in the world. Karachi swelled to 10 million inhabitants as environmental refugees flocked to the city, an exodus documented in Fred Pearce's book *When the rivers run dry*. These lawless slums became a breeding ground for Al Qaeda. Water is life and a lack of it leads to war in short order. Another example is the recent Syrian war. Its originals can be traced to the worst drought in the history of the country, a disaster that saw millions flow to urban slums, but precious little support came from the country's leadership. This happened right before the Arab Spring and the arrival of Islamic fighters – who found a fertile recruiting ground among the millions driven to despair. War and water are connected by bright red lines. Pulitzer prize winning journalist Thomas Friedman has analyzed the links, which have been recognized by former US Secretary of State Condoleezza Rice and President of the European Commission Jean-Claude Juncker.

One of the things I learn from all this is that between water shortages and ecocide or even war is a strong community. But in my experience, that is also no guarantee for success. When the ongoing climate breakdown wreaks havoc, even a model village can be swept away in minutes. The world no longer consists of a series of oases like Karimabad. In my book *Nepal, New pathways in the Himalayas*, I highlight the growing number of ticking time bombs climate change has created in the Himalayas. Growing glacial lakes there will one day swell to bursting point with meltwater driven by climate change, disasters that will sweep villages and fields downstream. In some valleys, as I learned while walking thousands of kilometers through the Himalayas, that has already happened. The most resilient of communities are no match for a tsunami. And that's why I'm convinced that resistance to the causes of disruption need to be both small and big: change needs to happen both at the local and the global level.

Soil: When the ground beneath our feet disappears

By sheer weight, humanity consumes ten times more earth materials now than it did a century ago. Exports have multiplied twelvefold, a faster rate than the economy as a whole.[116] Transporting such a massive amount of goods from one world region to another has led to some serious exploitation. Columbus and King Leopold were small beer compared to what is taken from others today. Stocks of essential but finite resources are dwindling. Who still believes that most trade transactions are fair, that information is evenly distributed and that the price of a product corresponds to the true cost of production, warts and all?

The number of states declaring war on each other seems to reach historical lows, but conflicts for scarce commodities have increased. Depletion of soil, water and biomass in the tropics happens under the guise of development aid. While that doesn't fit in the standard definition of war, it is a global attack on the ecosystems we depend on – in true Shock and Awe style. The white man's burden tries to transform this attack into some benevolent act and the labels greenwash ecocides. In reality, all these processes together form an unparalleled net transfer of biomass from non-industrialized to industrialized countries. This is not an opinion or ideological point of view, but a calculation. Andreas Mayer and Willi Haas, to name just two, quantified the material flows on earth by looking at everything that has weight and is traded: raw materials, fuels, biomass etc. They concluded that from 1950 to 2010, North America, Europe and China used significantly more resources than they extracted from their own environments.[117]

There's a pattern emerging. Industrial tree plantations mainly robbed the global south of water and smashed its biodiversity to meet demand in the global north.[118] It is an identical story for food. The *Patterns of global biomass trade* report states that almost all new agricultural land since 1986 has been for food exports.[119] Trade in agricultural products has grown three times faster than agricultural production itself. This despite the fact that yields have grown rapidly, particularly in sub-Saharan Africa. If we eat much more than we did in 1950, it is not because we produce more in Europe, but because our farms have sprawled into other parts of the world, flattening forests and churning up landscapes. Today, 40 percent of the food that Europeans eat comes from a non-European soil.[120] No other continent has grown so dependent on food imports.

The result is a form of imperialism, as I have already flagged. But besides that, there is a more chilling consequence still. Nature cannot keep up with demand. Consumption of food, wood and fuel has doubled in the last century. The human appropriation of net primary production – so of all biomass that nature makes – went from 13 to 25 percent.[121] By the middle of the twenty-first century, humans could take 44 percent before any other living being can take it, if we continue to invest in biofuels. If we undercut the material needs of other lifeforms on this planet, you would expect a decline in those species. That is exactly what is happening. Scientists speak of a sixth mass extinction of all life on earth. We are literally gobbling up the food of all the other animals, like a swarm of hedgehogs taking the cat's food while the cat stands by, watching and hoping something will be left for him. What scientists also say is that once a certain threshold of taking the food from other animals is exceeded, the food chain rapidly collapses, with severe to deathly consequences from the smallest amoeba to the top predator. During the fifth mass extinction, the top dog was the T-Rex. Now, at the sixth, it is us.

* * *

Flemish people are increasingly hungry. They now buy 2.8 kilos of food each day, and rising.[122] Yet half of it is not eaten. This is partly due to losses in the long chain from some far away farmer to the local supermarket. But it is also because we throw away more than 89 kilos of food a year. In the Czech Republic and Slovakia, it is only 25 kilos per year.[123] Tristram Stuart explains in his TED lecture how "The global food waste scandal" came about: in the ever-lengthening chain from field to fork, inefficiencies arise, he explains. Flemish people really depend on food imports. Without imports, Flemish stomachs would moan and groan like an emptying sink. Should the transfers to Flanders stop on some black Monday, a run on the supermarkets would occur. In normal circumstances, a supermarket would be empty in a few days, but during a panic it will be emptied in a few hours. Hard to imagine, but there would be food riots. But with agricultural area falling in Flanders and food consumption still rising, the dependency continues to grow.

Popular ignorance is occasionally illustrated and even encouraged in the media. An op-ed by farming syndicate leader Hendrik Vandamme explained why he is sticking to his diet of delicious Belgian steak. Climate activists are urban hipsters, he railed before applauding Belgian farmers for applying the circular economy. He neglected to mention that half the country's cattle feed is foreign and not rarely comes from where a part of the Amazon rainforest used to be.[124]

Climate change has nothing to do with being "hip". It is outrageous that the Trump administration is censoring the use of the words climate change in relation to agriculture. The conservative IPCC is unambiguous about our climate challenge. If nothing is done, this century we will see an implosion of agriculture. And because the IPCC lags behind the facts for political reasons, it is much more likely that this will happen before the middle of this century. And it's not just crops that need time to adapt to a changing climate. "Some people say we can adapt with the help of technology. But that's a belief system, it's not based on facts. There is no convincing evidence that a large mammal with a body temperature of 37°C will be able to evolve so rapidly. Insects can do that, but not people." I quote Professor Will Steffen, the lead author of two studies on planetary ecosystem boundaries.[125] In fact, dozens of scientists worked on Steffen's studies for 5 years, and Steffen was allowed to present them to the global elite at the World Economic Forum in Davos, Switzerland. Seven times, no less. His groundbreaking study published in Science says that we are breaking the limits of what the ecosystems we as humans depend on can cope with at four levels: climate change, biodiversity, biochemical flows and land use change.[126] In these four domains, we have already strayed dangerously out of the safe zone.

According to the Food and Agriculture organization, our farmland is good for another 60 years.[127] The number 60 cropped up again when researchers looked at the intensification of agriculture in the last century. Soil erosion has accelerated 60-fold.[128] Only a generation ignorant of scarcity could be comfortable with the sword of Damocles hovering over our heads. When push comes to shove, mankind is little more than an upright chimp nurtured by a thin atmosphere above and an even thinner band of soil beneath. It needs both.

There is a report with a long but relevant title: *Wake up before it's too late. Make agriculture truly sustainable now for food security in a changing climate.* I will sum it up in one sentence: "Small farms are the only way to feed the world."[129] The authors are not pastoralists with goat-wool socks, but a committee of scientists brought together by the UN's Trade and Development Agency. Another study showed that people who work small vegetable gardens produce four to 11 times more food per hectare than large-scale farmers.[130] Monsanto and the Farmer's Syndicate want us to believe that only genetically modified crops and large-scale agriculture will drive hunger out of the world. The reverse is true: these methods cause land grabbing and soil depletion, making poor people go hungry. I learned how small farms are the future, by joining one.

Common sense farming

Tom Troonbeeckx is a farmer who occasionally mingles with top level CEOs, at home and abroad, to explain his miracle farm. His story has inspired thousands. At first, Tom did not see any future in farming: "Starting a normal agro-company requires too high an investment and too high risks." But a cycling tour through Eastern Europe aroused his love of growing food to levels that made him think outside of the box. "In Hungary and Romania, there is still an agricultural culture, unlike the cultural desert created by the rise of industrial agriculture." During Tom's bio-farm studies, a visit to a special farm in the Netherlands inspired him.

Tom: "I have a green-minded mother and a businessman as a father. The perfect genetic pool to start a biological farm." Tom started by buying a small old cherry orchard in Belgium with a piece of land next to it. The Open Field became Belgium's first Community Supported Agriculture farm or CSA farm. Participants, numbering 50 at first, paid Tom once a year for a share of his produce, which the members harvest themselves.

Tom's one-man business soon ramped up to 320 clients and a long waiting list. Other CSAs started to emerge, clearly inspired by Tom. Around the provincial city of Leuven there are already five such outfits and over 40 in Belgium.

Tom feeds 320 people with his 1.5 hectares of fertile soil. The risks of poor harvest are spread across all participants, making farming less risky. But the participants also win a lot. They can harvest over 100 different vegetables. There are no tomatoes year-round, but once the tomato season is there, you will find up to ten varieties, from red to yellow. And yes, you can taste the difference with the tomatoes from the supermarket. Everything is organic and as fresh as it is possible to be. There will be no fossil fuel for transport, cooling or for plastic packaging. There is no parking for cars, so people come on foot or by bike.

Compare this to the four barrels of oil per year that the average citizen needs to deliver him his food and drink. That is what it takes to fuel and make agricultural machinery, fertilizers, pesticides, packaging, truck cooling, supermarket heating, the plastic of shopping carts and finally the drive home. Now imagine a world where that fuel no longer exists.

Tom stores more carbon in the ground than he and his 320 customers emit, thanks to his cultivation method. He also maintains seed diversity by adding some old and almost extinct local crops to the mix. That is not unimportant when you consider that between 1903 and 1983, 93 percent of seed species were abandoned and lost forever, thanks to our changing buying habits.[131]

But there is more. The Open Field is not only an ecological, but also a social and democratic project. On that front, we can choose our contribution to this sweet deal within a price fork, and with an average contribution needed to run the farm as a guideline. Tom leaves it to us to consider if we find ourselves below average, average or above average earners. So we decide how much we pay for his work. This is so hard to fathom that I've had to confirm to the editors of this book that this really is the case. In addition, members of the farm decide about Tom's income and the farm's main policies. His income depends on how happy his customers are, people he meets on a daily basis. Compare this to farmers dealing with big supermarket chains who rarely give a hoot.

On the social side: Tom also works with school drop-outs, kids in need of time-out from the formal education system. For them, it is a chance to be outside working with their hands, away from the pressure and the high expectations that society imposes on them. And then there are the memorable evenings during the three annual festivities with delicious local food, music, a campfire and a harvest that money cannot buy: a sense of community.

Like in Hiware Bazar, the key is to think long-term, ecologically holistic and about community building. It is a total revolution in the raw material, transport, production, consumption to waste lifecycle, a revolution that goes miles beyond the exotic vegetables with organic labels that you can buy in a plastic bag in a supermarket.

This best practice is a welcome relief from all the bad news in this chapter, so allow me to wax on. Tom also keeps bees. Mainly to nail the pollination of his fruit and veg, but also for the delicious honey. A few sheep and cows provide necessary grazing and fertilizing. Rainwater is harvested for a modest scale irrigation system. The Open Field is a vibrant, local circular economy. If the world economy suffered a heart attack tomorrow, it would have little effect on the healthy food supply of the 320 people who eat from a 1.5-hectare plot within cycling distance of their homes.

Tom is particularly fond of the transparency and directness of his food chain; meeting the people that eat as a result of his hard work. "It's really encouraging to hear a nice story from time to time, for example from someone who says that his children are eating vegetables now that they come from this field. If you work for the supermarket chains, they will never give an encouraging word and will only want to get the price down."

This short food chain had another advantage. I became a member when my first daughter turned 2. That is just old enough for strawberries and raspberries to follow the shortest chain possible: plant to mouth. By the time she turned 5, Flora understood that something was wrong with kids-TV-star-branded strawberries shipped in plastic tubs to supermarkets in the middle of January. Tom's farm is sometimes cold and wet, but for our family it has become something of a spiritual place. Our weekly visits feel like a mini-pilgrimage. Whenever I return, I feel blessed and light on my feet. I can imagine experiencing emotions similar to many Sunday church goers.

* * *

Here's one last thing to consider regarding CSA farms. On 15 September 2008, the day Lehman Brothers went bust, bankers made frantic calls to their wives instructing them to drop everything and stock up on supermarket supplies. No joke. Joris Luyendijk recorded several such confessions by London bankers who were convinced supply chains would grind to a halt within hours.

In 40 or 60 years we may well experience a crash in industrial agriculture, brought down by a lack of oil and soil. But long before that, the risk of an acute shock to the system increases daily from unfettered capitalism based on a fragile financial system. In these conditions, the smartest thing to do is to jump on the CSA bandwagon before such a crash happens.

Chapter 4

How we take what the earth gives us

From Au (gold) to Zn (zinc) and from tree to soil, Part I was about what we as humanity take from the earth. We not only take more than we need, we take more than we can afford. Consumption of what we take rises way faster than population. The "we" in all this hides much more complex and worrying class division trends, as well as historic geographical shifts. Always more violence seems to accompany the unaffordable part of extraction, from underground chemical warfare to squeeze gas out of the earth to above ground warfare against earth defenders. The inequality and injustice we are creating are both growing and ever more hidden from view, especially for those on the pampered end that are shielded from reality by phony ethical labels. In cartoons, the EU is oft portrayed as a fragile lady upholding justice but through my journey I found that outside the EU she's not that different from the American eagle or Donald Trump. She grabs uranium in Niger to fuel French and Belgian nuclear power plants, destroys Indonesia for our shampoo and chocolates or chops trees in the Amazon to fire EU stoves. The EU is acting more like a bitch than a lady.

This uncomfortable reality of modern existence is usually hidden from us and when some of it makes the news, blame is quickly shifted to others – often the victims themselves. But exporting pollution to "under-polluted" countries and grabbing their resources on scales never seen before is the bedrock of the economic "success" of the post-war period.

Outside this reality there is a parallel universe that is also too little known. People like Bruno, Julio, Alexandru, Roger, Sumaira, David, Tom and millions of others are very busy resisting and forging alternatives. They are busy with building a future where organized citizens win power back at the cost of private companies and states. This is often referred to as the commons economy. On top of that, I see a degree of cooperation emerging between previously isolated small groups of earth defenders. They form a brand-new kind of multinational, international groups of resistance. Some political scientists are calling this the global movement for environmental justice.

Both the stories of destruction and those of positive alternatives are hidden but real and valuable. If you only look at the destructive story, you risk a point of total inaction, lethargy, cynicism or depression. If you focus only on the nice but small alternatives to run-away capitalism, a naive bliss will take hold of you. That may look like the most attractive choice, but given the urgency to change ways, we simply can't afford too many people doing that.

The choice is not black or white. A bit of schizophrenia seems necessary. Not just for understanding the world but also for looking at your own life. Most days I will eat a banana or something with chocolate from Africa. I probably use some palm oil from Indonesia now and then and coffee from God knows where, but certainly not nearby. My office printer runs on FSC paper. I use uranium imported from Niger, transformed by nuclear power plants into the electricity that powers my train. I have made more flights in my life than a fair share would allow me to. Call me a hypocrite, but I can and will not shut my mouth about our unsustainable existence. If only the Gandhis and modern hermits have the right to speak about our environmental problems, it is gonna be damn quiet on our collective walk towards the abyss.

These contradictions will not surprise the psychologists among us. They know that behavior and knowledge are miles apart. Harald Welzer wrote fascinating things about mass psychology, both about Holocaust denial by Germans in its midst, and the popular denial of a climate breakdown and the sixth mass extinction happening all around us in the present day. According to Welzer, it is totally absurd to think that people's thinking and actions align:

> Ninety-five percent of what we do is also embedded in the infrastructure of our daily habits and routines. We do not think about it anymore. There are not always the circumstances to behave in accordance with our conviction because we have several functions to fulfill. I have to answer different requirements for my profession, others for my family and still others for my relaxation.

I am with Welzer in believing that if we want a truly big change, we cannot just count on educating people with knowledge. If that only impacts 5 percent of what we do, it's just not effective enough.

That said, knowledge needs to get people to actions and solutions. It should be seen as a starting point, a means to an end. Avoiding denial and keeping our eyes and ears open to unpleasant truths is a mental battle that is probably easier to take if you also know some of the solutions to the many environmental problems. It's for that reason that I'll end this book with solutions. But maybe more important than being open to knowledge is being open to new experiences. Knowledge is surface level stuff for our brains, while strong emotions get deeper. Experiences often stir those deeper emotions, which are much more capable of bringing about change. So when I am asked, in a lecture or by readers, what they can do, I usually answer by telling them to just DO something. Do something you feel strongly about. Get a ball rolling. Or a spool. I once rolled a wooden spool two meters in diameter through the streets of Leuven. It was just rotting away on the pavement. Countless times I cycled passed it, thinking how it could be a great garden table. In the end I did not make a calculation about how many trees I was saving from the axe, I just jumped the mental hurdle of rolling the spool through the streets. The knowledge made me act but only the emotions kept the spool rolling. This turned out to be fun not just for me, but for those smiling at me as I passed them by. As David Dene said: "We're going to fight and we're going to enjoy every single step of it."

Resisting the destructive forces of capitalism and consumerism starts with a mental leap that anyone can make. Leap over the oppressive harness of societal norms. In the end, your biggest enemy is that part of your own brain that keeps you from taking bold actions for the greater good. We all need to leap over that fence in our mind. That is what the mothers from Greece to Balcombe did, what one Romanian antiquarian did or what I did by penning a simple email to my bank. Just start somewhere. The more people discover how exciting, fun and liberating it is to participate in this great transition, the more everything will snowball. History is knocking on the door. Will you open?

Chapter 1

At the end of the pipeline

Part one ended with the idea that experiences may be more important than knowledge. At the same time, our negative experiences are minimized by displacing the pollution associated with our economic success. We have not solved our environmental problems but exported them to what folks in the World Bank used to call "under-polluted" countries. Our mines have moved and more wood and food now come from Farawayistan than ever before. But we are also shipping back record amounts of our waste. We dump our aging ships in India, our computers in Ghana and we even send the shit from millions of our livestock to Africa.

The odd unintended side-effect of all this is that we deny a generation the opportunity to experience the shocking and harmful cause-impact relationships of our economic success. What the rich West does is what I did as a student in my university dorm room. I swept the dirt under the carpet. If you cannot see it, it is not there. It is about time we in the West became responsible adults.

But we haven't exported everything. Just make a "toxic tour" around Naples or go to Megali Panagia and the Skouries mine. In fact, for some pollutants my compatriots in Belgium do not need to go anywhere. Just take a deep breath on the streets of Antwerp or Brussels. Disclaimer: any harm, injury or death from doing so is not the responsibility of this author.

Chapter 2

The trade in hot air

1997. That was the year that countries decided they could trade their air. The Kyoto agreement was where the fun began, blowing hot air into a giant balloon now worth more than €100 billion a year. The carbon part of our air that the previously discussed plantation in Uganda transformed into biomass is part of this new global industry. It was dreamed up by people like Al Gore who, at the eleventh hour, worked it into the Kyoto Agreement. The first global agreement to collectively do something on climate change needed a so-called Clean Development Mechanism (CDM), Gore argued. Without it, the US would walk. Why? Because this meant the US could carry on as normal, not curbing emissions at home. The American way of life is not up for discussion, negotiators had growled.

So here we are now. Developed countries can buy "cleaned air" from developing countries. The deal sounds reasonable enough to the economist sitting in his desk: it is cheaper to reduce emissions or turn airborne carbon into biomass carbon in countries with less expensive workforces. One of many glaring flaws though is that carbon infested air is still killing those in the West. The 10,000 Belgians who died of dirty air last year can of course no longer complain about their dirty air. Same story for the 400,000 Europeans whose deaths last year were hastened by dirty air.

But the economist will reply: well at least people in developing countries will be able to take deep breaths of cleaned air. We already know that industrial tree plantations are displacing people, something unlikely to improve their health. Another carbon credit project, this time in Mozambique, boils down to modern slavery.[132] Farmers there are forced to sign a contract to maintain a tree plantation for 7 years at $63 per family, per year. For the remaining 92 years of the plantation's productive life, they have to maintain it. For free. A pioneering form of intergenerational slavery? The company that organized this project in Mozambique, Envirotrade, was founded by a man on the run from the authorities after being convicted of training a paramilitary group. The same man had financial ties with Chile's brutal late dictator, Augusto Pinochet. Yet this was a project heralded by the UN at the Rio+20 summit as a shining example of the fight against climate change.

So what goes on below the surface of those projects deemed unfit for the limelight? Carbon credit projects in Africa are a new form of colonial slavery, according to an international team of academics led by Professor Patrick Bond.[133] These so-called climate mitigation projects are, in their opinion, just another way to grab farmlands from peasants. That is one characteristic that most major development projects have in common: they rob land and independence from the most vulnerable people. This sleight of hand has simply shifted the name and the blame from colonialism to clean development. It is their fault if they do not grab the "opportunities" we are kind enough to offer, goes the thinking. In the West we neglect that the real "they" is all too often some crackpot dictator that does not give a damn about most people he (usually it is still a he) reigns over, and with whom our companies and countries make deals that displace locals.

* * *

In October 2009, I traveled to India to visit a massive paper mill that I would have been paying money to, through the Belgian government's plan to pay them to emit less carbon. The Indian company Yash Paper Industry claims it can help Belgium to fulfill its promises made in the Kyoto agreement. In exchange for my tax money they will reduce emissions in our place. Belgium's climate czar admitted that his administration can't visit the factories that deliver this great service to us but he didn't mind if I went to do a little on-site check.

Once in their region, Yash of course opens a charm offensive. A driver picks me up at the hotel for my first meeting and before I know it, the hotel's bill is already settled. On arrival, the firm's financial manager, Santosh, asks if I noticed anything amiss with my hotel power supply. I sure did, it had a habit of cutting out. "Then you immediately understand why we as a paper mill never use power from the grid," he said. Makes sense, I thought. But what didn't make sense was that by saying that, Santosh had already shot himself in the foot. The Belgian government was stumping up cash to help Yash go off-grid. A grid dominated by coal power production. By going away from the grid, emissions would be reduced and for that service, we Belgians would pay Yash Paper. But if Yash would not touch the grid with a barge pole anyhow, what are we Belgians paying for exactly?

I was guided around the biomass boilers that power the factory. Santosh says it has always run on this kind of green electricity. That I already knew from the company's annual reports, which include data on the levels of fossil fuel consumption. The great thing about Yash Paper is that they have always been close to zero. But then how can they get our money to reduce what they don't burn?

The whole carbon credit system is filled to the brim with jargon crafted simply in order to mask weak arguments and obvious flaws. The good news is that you don't need to juggle with their jargon to understand you cannot reduce production of something that you're not making. The climate does not tolerate magic tricks. Only humans are capable of fooling themselves in such ways.

To get this trick done, you need a financial expert who can hide fraud a bit better than Santosh. But I wanted to try and play this game according to their rules. I was not satisfied with taking apart this obvious fiddle with simple logic alone. I wanted a smoking gun and one that made a mockery of their own jargon and bookkeeping magic tricks. An Indian specialist for whom the Indian bureaucracy has no secrets tracked company documents in Hindi held by numerous Indian authorities, including the Securities and Exchange Board of India. It was soon very clear that Belgian funds were never going to reduce greenhouse gas emissions at Yash.

I will spare you most of the details, but one of the six magic tricks the firm had pulled stands out. The trick is called barrier inflation. Yash simply added hot air when reporting the price it paid for its boiler to a level well above the usual market price. This made it appear that Yash needed a subsidy to be able to install a green energy boiler and thus avoid the electricity from that grid (which they wouldn't take anyway). My story, The Big Carbon Fraud, appeared in MO* Magazine in 2009 and later also in India. Shortly afterwards the opposition Green Party invited me to explain the case in detail so that they could shame the relevant minister. After all, Belgian taxpayers spend over €100 million subsidizing industrialists in exotic countries who go on to use a small fraction of that money to provide foreign journalists with a delicious curry and hotel lodgings. Thanks, taxpayers! The rest of the money finds its way into the renovation of some fancy villa or on a private jet. Taxpayers have the right to know what they are paying for, not least when it is supposed to be spent combating perhaps the most serious threat facing humanity.

The Belgian climate administration at first only said that my findings "do not exactly match their data". But it admitted that additionality (read: authenticity or simply being real) is indeed "a very complex issue". It also admitted it does not have the time and resources to monitor each project on location. I can understand that. The next question is: was Yash an exception? Doctor Barbara Haye of the University of California, Berkeley examined 85 similar climate projects in India and found enough smoking guns to equip a small army. More than half were not reducing greenhouse gas emissions, she showed. Lowering the burden of proof to elimination by simple logic rather than hard evidence of fraud, the rate would rise to an astounding 90 percent. In all these cases, Western countries claimed they did their part to meet their Kyoto commitments. Unsurprisingly, climate emissions have continued to rise and fast.

This is unfortunately not just an issue in India. The whole CDM is built on system errors. As with wood and palm oil, a few companies are paid by factories to do the certification. Factories pay them and they want to get the credits, so why be more strict than the two or three other certification companies? That would not make business sense. The developing countries to which the money from developed countries flows should monitor the certification companies, but what is their incentive to punish them and in doing so harm the money flow to their companies? India never disapproved a single project. Why should they? Every Euro in the coffers of Indian businesses is welcome. And for the countries that give money, they do not really mind if they get what they paid for. For them, the system is just a fast and inexpensive way to maintain the status quo at home while claiming they did their bit to tackle catastrophic climate change.

In the case of Yash, Belgium ended up withholding its funds. As a journalist, I could claim credit for that, call it a cool result of my work. But I have no illusions: Yash probably sold their hot air to a less demanding country and Belgium just switched its payment to another factory, most probably equally fraudulent. Meanwhile, nobody benefits from the more smoke-free air we are supposed to be paying for, not in India and not in Belgium. In Belgium, life expectancy has fallen by a full year due to air pollution. In Antwerp, the figure rises to 3.5 years. That also costs us billions in healthcare each year.

The need for fossil-fuel fumes free air is so acute that some documentary filmmakers hatched a creative idea to claim a slice. So they hired a truck and drove to Hungary. Belgium had paid the Hungarian government €2 million in 2011 for air that had got cleaner when its economy shrank drastically after the fall of the Soviet Union. While the European Commission dragged Belgium into court in 2011 and 2015 over its poor air quality, Belgium found money to buy pure Hungarian air from 1991, without the satisfaction of breathing it in Belgium. Of course, the Hungarians just said thanks and used it to pay for anything but efforts to combat climate change. The filmmakers came home with a truck full of Hungarian bubble plastic, dropped it at the office of our climate minister with a post-it on it, reading: "Here's the first installment of the clean air we paid for!"

Bad as this already is, there is worse to come. In Belgium, the federal state that was planning to pay Yash actually does make an exceptional effort to prevent buying air castles. They invite NGOs to choose their criteria for granting CDM projects a subsidy and they support journalists who want to go into the field and nose around. The system failures are not the fault of the Belgian administrators, who do their best to execute political orders as best they can. But none of these extra cautions exists at the Flemish regional level. Flemish Minister of the Environment Joke Schauvlieghe proposed in 2009 to buy more of these carbon credits. *Flanders purchases 20 million tonnes of clean air* was the headline in all the newspapers. The minister explained that this was a great opportunity, because the market price was so low.[134] Never did the journalists wonder why the price was so low.

I will tell you why the price was ridiculously low. There are factories in India and elsewhere where the production line stands perfectly still, but it still rakes in profits. They essentially threaten to make something very polluting and then accept cash not to. This state of affairs fosters a culture of sanctioned blackmail and entitlement. It converts productivity into idleness in countries that would be better off working to advance themselves. Oh and this house of cards costs all of us more than a €100 billion a year.

My former colleague Sarah Vaes once managed to get the entire fraud on the front page of a major newspaper.[135] This caused insufficient electoral problems to change matters. The climatological bunkum seems too lucrative and opposing voices too weak. Almost all politicians continue to defend the status quo. Even the Greens dare not grasp the bull by the horns and are content to amend rather than replace the system, a tack they have been following for 10 years now, without any significant success. Pressure groups and concerned scientists have reached more forceful conclusions; the whole carbon trade should be recognized for what it is, a massive failure.[136] This is evidenced, for example, by a petition signed by more than 140 organizations to "Scrap the ETS" – the European section of the carbon supermarket.[137] Or by the success of the book The Myth of the Green Economy by researchers Anneleen Kenis and Matthias Lievens.[138] So far humanity has lost 10 years and a stack of cash, and we are not even treading water. And we are still coughing our lungs out at home. As with phony ethical labels, large NGOs are afraid to shoot down the CDM and carbon trading in general. There could be several reasons for this. Maybe it would cause funding lines to dry up. Maybe they prefer to toe the line, within what is politically acceptable today. Or maybe because pulling the plug on something you have backed for years would involve eating too much humble pie. Last but not least, it could be because an unshakable faith in a free market solution to everything tends to prevail even within the environmental movement.

Chapter 3

Sustainable destruction

In a bend of the River Schelde at Hoboken in Belgium, stands a factory that has been belching smoke since 1887. Name an element on Mendeljev's periodic table and it is likely to have passed through this factory's gates at some stage. Until 1960, it was heavily involved in processing raw materials from the Belgian Congo. The chimneys are now owned by Umicore, a company that turned around its colonial reputation in part after winning a prestigious award as the most sustainable company in the world in 2013. My conversation with the boss, Thomas Leysen, was on a less happy subject. I was there to present him with a long overdue invoice for €312 million.

Before I get to the invoice, there is one more thing to say about that corporate knight's prize (named after the Canadian research institute that hands it out). The criteria for winning it include "financial health and recent payouts in court cases". Sound odd? This sure is not the first thing that comes to mind when one thinks about the word sustainability. It is true that Umicore built the largest recycling unit for precious metals in the world in Hoboken. The company underwent a grand ecological modernization that began in the 1970s and reached a high-water mark in 2004. That year, while the Belgian Green Party was in power, Umicore shelled out €77 million to clean up the surroundings of two factories it had operated on, one in Hoboken and one in Olen. Most people in Hoboken were satisfied. "They even replaced the ground under our garden shed," one old-timer told me.

But this is only one side of the coin. The cleaner side. The other side is one that some big NGOs do not much like to talk about. They work well with Umicore in all kinds of sustainability projects and do not want to rock the boat. I had to convince the leadership of some of Belgium's biggest NGOs that the study I was about to unleash was worth risking their cozy relationship with a company seen in Belgium as one of the good guys.

My study confirmed that the heritage of 130 years of pollution is still very much present. The closer you go to the plant, the higher the lead, arsenic and cadmium concentrations in the soil. The company's clean-up was limited to a much smaller perimeter than the range of pollution, and even there, only the upper soil layer had been removed. Besides that, a lot of damage had been done that remained unsettled.

In 1973, the local government sent a drastic letter to all residents of Hoboken warning them not to eat anything they might grow in their gardens. A cherished Belgian habit of growing fruit and veg was over; the soil too toxic. The move may sound over the top, but is not. There is undeniable evidence of a connection between the distance to the factory and the amount of lead in the blood of toddlers, according to the Provincial Institute of Hygiene, which has been researching it since 1978. The costs of this detection and the treatment are borne by the taxpayer, despite the fact that a private factory caused the mess. There is something called the Polluter Pays principle, which all countries signed up to back in 1992 at the first UN Rio Summit.

Umicore does not even deny they caused this increased lead in blood values or the fact that it polluted Hoboken's gardens. But it is not ready to admit a link with cancers. I managed to obtain detailed data from the Belgian Cancer Registry and the results were shocking. Cancers are significantly more common in Hoboken than elsewhere. Lung cancer is particularly prevalent. The number of lung cancers among women in Hoboken is twice as high as the Flanders average. That could still be written off as a coincidence. Yet I was also able to surface internal Umicore documents warning about the dangers of lung cancer.

I received the document through an employee who risked

losing their job by passing it to me. It literally states that lung cancer would be the most likely cancer resulting from their production process. This internal research was used to reassure the workers that the *other* forms of cancer they were experiencing and complaining about were not related to their work in the factory. The study was carried out by a professor paid by the company. He argued that the workers were well protected from lung cancer because they wore masks at all sensitive locations in the factory. That the same toxic substances spread out over an area where about 3000 people live was not in the scope of his study. But that does explain the massive numbers of lung cancers in the vicinity of the plant.

These connections were again confirmed through an extensive study made in the late 1990s, based on patient data from six doctors with patients in Hoboken. That study showed a clear positive correlation between the number of patients with cancer and how close they live to the plant. But the doctors from *Medicine for the People* did their research in their spare time, on top of their day jobs. They are not associated with a university and they are thus not connected to the peer review process. This means that Umicore found it easy to shoot down both that and my study, by saying that "popular epidemiology" and "post-normal science" are de facto worthless.

But the uncomfortable reality is that what the doctors found, what the figures of the Belgian Cancer Registry showed and what a professor paid by the company concluded all point in the same direction. In light of the levels of pollution, it is only normal to see such cancer rates. The levels mirrored other cases where toxic substances had led to a spike in cancer cases, as detailed in International Agency for Cancer Research reports from 1980, 1987, 1993 and 2006. The study I made was published in a handbook on ecological economics that is still used around the world today.[139] It got the full support of Professor Joan Martinez Alier, an *eminence grise* in the field of ecological economics. He personally supervised the research. The association for environmental professionals in Belgium dubbed it a groundbreaking study.[140] The scene was set for my meeting with the Umicore boss. The firm owed compensation to the tune of €312 million.

How did I reach this figure? We start with €12 million for the diagnosis and treatment of children in Hoboken with increased lead levels in their blood, using decades of records from the authorities who monitored this. Second, there is around €59 million that people in Hoboken have spent on vegetables and fruits that they would not have spent if they were allowed to continue using their gardens between the 1978 letter and 2008, the end of my study. Despite benchmarking these figures against as much relevant data as I could find, this amount will carry a relatively wide margin of error amounting to several million Euros. Third, treating additional cancer cases amounts to approximately €35 million, based on treatment costs recognized internationally. Finally, compensating the families of those killed by the company's pollution would amount to €206 million, again based on recognized rates with a jurisprudence.

Walking into Umicore's headquarters in Brussels to present it with my bombshell bill, I was flanked by a few companions: Professor Tom Bauler; Léa Sebastien, the PhD student that supported my research; and Leida Rijnhout, coordinator of the former Flemish Platform on Sustainable Development. We were meeting Thomas Leysen, Chairman of the Board of Directors.

Mr Leysen listened patiently, but made it clear that his lawyers would contest the link with cancer. But he seemed open to negotiate over the costs not related to cancer. He said he wanted to look in the mirror with a clear conscience. I believe that. We got the impression that Thomas was rather ashamed of the dirty legacy that he had inherited when he came to the helm of Umicore. He is active at the World Business Council for Sustainable Development and several years later I discovered that he is active in pushing KBC in a climate-friendly direction (see part I). But he did not want to put his business at a competitive disadvantage. At the end of the day, he answers to a board of directors and shareholders, not to the thousands of people who live around his factory. That is the system error.

Thomas Leysen: "I want to create a level playing field, so that everyone has to make the same efforts in terms of sustainability." That is indeed what should happen, but the level should be that the polluter pays, not the state or the local people.

The entire lead-in-blood story resurfaced in 2017. Blood tests of toddlers showed that lead was up to seven times over World Health organization limits. The cause was quickly identified as coming from Umicore. While replacing a roof, a lot of lead from the production process had escaped into the air over a period of 4 months. How the company failed to see this coming is beyond me. Perhaps Thomas would have been more careful in his orders if he had had to explain to the parents of Hannah and at least 14 other toddlers with increased lead levels that she was likely going to suffer from irreparable reduced concentration, headaches, inertia, anemia, movement problems and a lower IQ.[141]

The people in Hoboken can only hope that one day a good lawyer will take up their case. We must all hope that some enlightened policymaker creates the level playing field that Thomas Leysen talks about and that the level is where it should be. I am not a lawyer, nor a policymaker. My involvement with Umicore stopped right at the point we presented our findings to anyone who would listen. We went to talk with a group of environmental lawyers in Ghent, but the small NGO we worked for did not have the budget to pay for a legal case.

Almost no major company would make a profit if the full costs of production were paid by them.[142] This is the shocking finding of a comprehensive study by the United Nations Economics of Ecosystems and Biodiversity program. It showed that we consumers are living in a dream world where the costs of products decline while hidden costs paid for by society go up. Costs for things like cleaning up pollution or providing healthcare. And it is precisely because of that process that conflicts to protect the common good seem to be on the rise.

Any economist will say that internalization of external costs is required. But it seems that most of them neglect that ever more costs are externalized After all, the more important question is: how do you do that? A profit-making entity is not going to do that by itself. It needs to be controlled. A strong government needs to create that level playing field so that responsible companies are not put out of business by those that continue to externalize costs. A strong government must regulate, control and police. Unfortunately, recent decades have seen the opposite trend: deregulate, privative and provide one fiscal amnesty after another.

Chapter 4

Planned obsolescence

What companies get up to when there is little regulation and no strong government oversight goes beyond what most people think. I am not one for conspiracy theories, but some are true and the planned failure of products by their designers is one of them.

The first nylon socks were so strong that you could pull a car with them. That was a problem for manufacturers. Once everyone had two pairs of nylon socks, no more socks would be sold. And so Dupont's product developers were instructed to make the nylon less firm so that it tore quicker. This is not a conspiracy theory, but a well-documented fact. The same forces are at work across a large part of manufacturing. The logic is pretty simple: the sooner a person bins a product, the sooner they become a customer again. And it did not take much time to catch on. As early as 1932, Bernard London wrote the paper *Ending the depression through planned obsolescence.* What started with nylon socks expanded to cars, textbooks, washing machines, microwaves, software, mobile phones, laptops, stereo kits, coffee machines, displays...the list goes on.

The numbers are interesting. To make a computer and monitor you need 240 kilograms of fossil fuel, 22 kilos of chemicals and 1500 liters of water.[143] People may be surprised to learn that computers take a huge amount more energy to make than they will ever consume during their use. "To equal the energy used for the production of a laptop, you have to use that for more than twenty years," according to Carsten Wachholz, a former product policy officer at the EEB. Yet computer lifespans are getting shorter. Ones made in 2010 have a 10 percent shorter lifespan than those made in the year 2000[144]. As a result, electronic waste has grown to more than 50 million tonnes per year.[145] Consumers globally have bought more than 7 billion smartphones since 2007.[146] Electronic waste represents just 2 percent of the waste in American dumps, but makes up more than 70 percent of the total amount of toxic waste.[147] Approximately 80 percent is exported to Asia, where millions of people suffer from severe health damages caused by e-waste that burns and leaks into air, soil and water. Electronic waste is the fastest growing waste stream in the EU, accounting for around 10 million tonnes a year. In Flanders, the waste flow increased from 8 to 22 kilos per person in the decade from 2006.[148] Growth has not been a steady linear increase, but a raging dumpster fire characterized by exponential growth. That fire is spreading to developing nations. The amount of e-waste could rise fivefold in countries like India.[149]

But how do you plan for obsolescence? Each self-respecting research department now has robots, according to Lode de Geyter, general director of the West Flanders University and a former electrical engineer. These robots push buttons thousands of times and test devices in a variety of ways until they find weaknesses. At this point they should be figuring out how to avoid a breakdown, but instead they calculate whether failure will likely occur before or after the warranty period is over. One trick is to set heat-sensitive components right next to internal heat sources. Hey presto, before long you have a broken product. Another trick is to hide malicious chips deep inside a printer. On some idle Tuesday morning these will switch the printer into a state of non-cooperation and their owners into a state of rage. Instead of asking to replace or clean the ink cartridge, the printer will suddenly tell us that a repair is needed.[150] A true "Printer says no" moment, with no solutions offered. It is a piece of cake to then make the repair costs more expensive than buying a new one.

* * *

The problem is not the technology, but the capitalist economic system. Scheduled aging is not limited to electronics. Publishers update textbooks on subjects that have not changed, reordering the information and encouraging students to buy the shiny new version.

Thankfully, consumers have clocked what is going on and formed strategies against this buy-new culture. Student groups developed guidelines to keep track of courses using the perfectly good older textbooks. Or take Tim Hicks. Tim fixes laptops. On his website, there was a manual in PDF for each model, until Toshiba lawyers forced him to remove the manuals for more than 300 Toshiba laptops. Toshiba knows that if manuals are not online, there is no alternative for the consumer than returning the defective device to a pricey service center that will tell you "it's going to take a few weeks and cost you a lot but we have a new version on promotion now!"

Disgusted with all this, people have built a blossoming network of fixers and repairers.[151] A YouTube video shows how to get that chip deep in your printer out and the machine working again.[152] There is a "repair manifesto" and a whole subculture of activists with revolutionary slogans under the credo "If you can't fix it, you don't own it". People also unite in repair cafes where do-it-yourself volunteers are happy to repair your stuff for free.[153] The French branch of Friends of the Earth created a website full of practical tips to make products last longer.[154] German economist Stefan Schridde became famous with his website Murks, Nein Danke! He has solutions for planned aging in a broad range of products all available in seven languages.[155]

But we need to really pull the plug on planned failure. The leasing sector is expanding and today a company like Philips offers Lifecycle Services for lighting systems. You do not buy lamps and bulbs, you buy a fancy installation giving you sophisticated lighting where you want it. Philips installs but still owns the hardware so suddenly has an incentive to ensure its kit lasts as long as possible. For Philips, this is a profitable way to go back to the old days when they made bulbs that lasted for over 100 years. Consumers are not conned and benefit from a high-quality setup.

Most governments are only slowly waking up to the problem of predatory businesses engaging in planned aging The Austrians got ahead of the game in 2015 by launching a new label to reward durable, repair-friendly electrical and electronic equipment. France took a more forceful approach by creating rules against planned aging that could jail company executives for up to 2 years and slap on fines of up to €300,000.[156] In Sweden, lawmakers halved taxes on repairs for apparel. Most governments are only slowly waking up to the problem of predatory businesses engaging in planned aging.

Broadly speaking, it is NGOs that go toe to toe against companies, as well as pressure valve resistance movements by people such as Tim Hicks, Stefan Schridde and Michel Bauwens of the Peer-to-Peer Foundation. This foundation brings people together around things they make and use themselves, freed from the profit motive that steers capitalism. Wikipedia is a peer-to-peer product that has now totally eclipsed its capitalist counterpart, the Encyclopedia Britannica.

What is still missing is a stronger European and global approach. The European Economic and Social Committee (EESC) has slated some good ideas, including that producers should shoulder recycling costs if their products last less than 5 years, and that the minimum lifespan of a product should be stated on its label. It also advocates for spare parts to be made available for at least 5 years even if manufacturers end the product line. Ending "industrial dictatorship" is what Jean-Pierre Haber calls his work as member and co-rapporteur of the Advisory Committee on Industrial Reconversion within the EESC. Thierry Libaert, also EESC member and rapporteur, tried a similar tack: "The Committee supports a complete ban on defective products built to shorten its lifespan." But the EESC is merely an advisory body. Groups such as the European Consumer Protection organization (BEUC) do their best, but they enjoy no formal power. Meanwhile, the rise of populist Eurosceptic voices makes bigwig political leaders in Brussels look for ways to sideline eco-design regulations instead of strengthen them. The prevailing mood is that regulation is bad, no matter what it actually does.

Chapter 5

Circular economy?

In 2014 I joined the EEB, the largest federation of environmental NGOs in Europe. My new colleagues were delighted with a package of product and waste measures that the European Commission had published. We were going to be moving towards a "circular economy". Among all the cheers it sometimes seemed like extraction and waste would soon become a thing of the past and that all the problems described in this book would cease to exist.

A little later, I was in Copenhagen listening to Professor Joan Martinez Alier explaining to experts from the European Environmental Agency that the industrial economy is not circular "and can never be circular". What was going on here?

Joan used the second law of thermodynamics to explain that you can use the energy in fossil fuels only once and that the whole economy, including the circular economy, depends on those fossil fuels. Today we use around 15 billion liters of oil. Every. Single. Day.

Who was right, my new colleagues or the professor I was working for? The answer depends on whether you see the EU package as a glass half empty or half full. On the upside, the package would generate 415 megatons of CO_2 savings and 870,000 additional jobs. But if, say, Umicore uses more fossil fuels to burn broken mobile phones to extract a few grams of gold from them, is that progress? Can the circular economy mean two things at once: a step forward in reducing extraction and creating jobs, but a step back in using even more fossil fuels? Whether the buzzword "circular economy" helps us forward or backwards depends entirely on the extent to which its implementation is holistic. Is it just about material flows or does it also include the energy factor? And what about the facts on resource extraction?

All of a sudden, the question of who was right became academic. Despite years of work and all the benefits that the package offered, a newly appointed European Commission under Jean-Claude Juncker decided in autumn 2014 to bin the lot. Juncker's team had a simple sheepish mantra: less rules good, more rules bad. And just like Orwell's sheep in Animal Farm that slogan applied to everything and everyone and is being repeated over and over again.

Professor and engineer Willi Haas and his team wondered how circular the global and European economy already are. Every year, about 4 gigaton of materials are recycled, while 62 gigaton of new material enters the economy, the scientists estimated. The world economy was thus about 7 percent circular.[157] The other 93 percent saw the digging up, cutting down and hauling about of finite raw materials.

For me, that study was an eye opener. It showed how far away we are from a true circle. The economy is still basically a massive pipeline with very small loops here and there. Haas and colleagues also calculated that about 44 percent of everything that is dug up is eventually burned and thus no longer available for circularity. Willi and company basically concluded that our economy is a pipeline so thick that you cannot bend it into a circle. Strategies that focus at the end of the pipeline cannot work because there is just too much flow. They advocate a switch to renewable energy, more drastic eco-design and a significant reduction in the amount of stuff we put into the gaping mouth of the pipe that we call "the economy". They argue there is no way around the need to dig up a factor less than we do now. This means economic degrowth. And sadly this is currently the stuff of political science fiction. The word degrowth itself is even taboo. Meanwhile, the global economy went from 62 to 78 Gigaton between 2005 and 2010. These levels put the idea of making the economy circular in the realm of fiction.

The gap between political and scientific realities is big and getting bigger. Take the Oxford scientists who at the start of 2017 argued that climate change is but one of many threats facing humanity and that the interconnection or interference of different trends of environmental breakdown is still underestimated.[158] In mathematics, one plus one equals two. But when two storms suddenly join forces, the net effect can be a whole lot bigger than the sum of its parts. The scientific team, which includes five top scientists from several national academies of science, points the finger of blame at the combination of growth in the use of raw materials and land, but also at forms of pollution other than atmospheric. In scientific language, they say that the conditions for maintaining human societies at their present scale are being undermined. In common parlance, they are saying we are fucked if we continue like this.

Meanwhile, we keep doing the opposite of what scientists say we have to do. Consumption of materials increases by 3.6 per cent per year on average.[159] The world population is increasing by 1 per cent per year, although that growth is slowing. The UN forecasts that the global population will grow by 20 percent by 2050, but consumption will grow by 200 percent. The key issue is clearly our overconsumption, not overpopulation.

Earth can sustain 10 billion people for a very long time to come. But not 10 billion Western-style consumers. That is why a real circular economy has to be part of our solution, while realizing it is not a silver bullet. It needs to come together with a plan for degrowth. Fortunately, Jean-Claude Juncker's deputy, Frans Timmermans, went to the trash bin to dig out the circular economy package his boss had binned in order to try and recycle it. The new version is not as good as the old, but my colleagues are working night and day to make it more ambitious. My job is to argue that you also need to keep resources in the ground in much more radical ways. Here's why: 71 percent of all resources circulating in our economy are not even recyclable – such as

Chapter 6

Toxic tour in mafia territory

Despite being experts in exporting our shit, so to speak, we have kept some of it in Europe. We can "thank" the Naples mafia for that. As part of my work, I did a toxic tour of their playground: the Campania region around Naples."If anyone comes, hide your cameras," warns the Italian activist as he leads us beyond a no-entry sign of a landfill in the region. "*Cinque minuti*", he adds as we get out of the minibus – we do not have long. We jump out and start shooting pictures like a busload of eager Japanese tourists. I zoom my camera in on a leaking battery then out to snap the mountains of waste. These mountains are made up of waste balls; a mix of toxic crap compressed into a ball a few meters in diameter. The consortium of companies who made them call them *ecoballs*. They either have a sense of humor or they are referring to the economic benefits. Either way, there's nothing ecological about these balls. Inside, they will stuff anything, from plastic bags to asbestos plates.

After a few more pictures, the leading local activist calls us back to the minibus. In the distance, a truck approaches. We are just outside the gate as it rumbles past. From its livery, it is supposedly transporting beer. But the activist tells us that he's simply delivering a fresh batch of *ecoballs*. Activists have mapped some 5200 illegal landfill sites in Campania by 2010. Campania is the forgotten sinkhole of Europe. Waste crime is one of the fastest growing forms of organized crime. Our skyrocketing e-waste problem is also fueling this dark economy, but you'll find all kinds of waste here.

Lucie Greyl has been studying the waste problem in Campania for years and she says that illegal managing is cheaper, so more convenient. Illegal trafficking of waste in Italy was earning the mafia somewhere around €7 billion in 2008, according to the Italian environmental organization Legambiente.[160] Like a leaking barrel, their profits had expanded to around €19 billion by 2015.[161] The authorities have been trying to grapple with this waste mafia empire, making 1602 arrests that year and dragging 871 companies into court.[162] It has had little effect. Italy holds the European record for "lost waste". Touring Campania, it is not hard to find it. The fallout is also pretty easy to trace. Residents here are four times more likely to get cancer than their compatriots.[163]

Another consequence is perverse: a circular economy for poison. We export it to Italy, the mafia sells it as compost and Italian farmers export their mozzarella and tomatoes back to us. Sounds unbelievable, but one researcher discovered that even nuclear waste from German power plants was being used as compost in Italy. Activists and journalists who uncover such stories risk their lives.

In 2018, I checked back on the fate of the *ecoballs* we photographed in 2009. "Well, they're still there", says Lucie Greyl. "At one point the regional waste plan planned to 'treat' and then burn part of them, but that was never implemented."

Despite the mafia's best efforts, the area around Naples is a Garden of Eden compared to Agbogbloshie in Ghana, the world's biggest e-waste dump. This hellish landscape comes complete with infernos of burning wires and circuit boards. The smoldering is actually smelting, fires tended by thousands of kids working to recover valuable metals. Many die from cancer before they are 30.

In Europe, only half of all e-waste is recycled. The rest is shipped mainly to landfills like this in West Africa and Asia. It is shipped there in disguised and illegal ways, because in theory the export of e-waste is prohibited by European law. But not every member state wants to tackle the problem themselves. A new law obliges these governments to make and publish accurate inspection plans. But free movement arrangements mean there is little to stop companies exporting their e-waste from the porous ports of Southern Italy.

Shipping hazardous waste outside the EU is a routine job for many companies. Only occasionally do things turn sour. Dutch company Trafigura delivered a gift to Cote d'Ivoire in August 2006: 20 containers of a super toxic substance. They had first tried to unload their toxic waste in Amsterdam but the smell was so foul that Amsterdam's environmental authorities forced them to stop and reload it. Then they went to Cote d'Ivoire, where through some shady dealings this stuff ended up contaminating the sidewalks of the poorer suburbs of the capital. The smell was so toxic that the Ivory Coast Anti-pollution Center later said it "may cause immediate death in the event of inhalation". But it took a while before the flooded hospitals knew what it was they were treating. The waste did kill at least 15 people and injured at least 30,000. The scandal forced Prime Minister Konan Banny to dissolve his entire cabinet.

So what just happened? Trafigura had calculated that it was cheaper to deliver deadly waste to some Cote d'Ivoire subcontractor who went on to do the actual dumping for them than to the port authorities in Europe, where it would have cost €1000 per cubic meter to properly process it. But in doing so, Trafigura broke the law and for once, it paid a reasonable price for its cold-blooded business model. It was forced to cough up $200 million in compensation to the Ivorian government followed by an undisclosed sum to the 30,000 Ivorians. But executives were never prosecuted by the competent authorities. No CEO had to contemplate jail time for unleashing chemical hell on tens of thousands.

The UN used the tenth anniversary of the incident to ask for renewed global attention to the violation of human rights it caused.[164] Thousands of victims still received nothing and much of the waste was never treated and is still hurting people in

Ivory Coast. Just a month after this "anniversary", the name of Trafigura reappears, connected this time with illegal shipping to West Africa of diesel a hundred times dirtier than that permitted under European standards.[165]

Chapter 7

Beaching boats in Bengal

There are other ways to get rid of European waste without the need of hiring a dodgy Ivorian subcontractor. Many companies simply load a rusting hulk up with toxic waste and run it aground on some forgotten stretch of Bangladeshi coastline. Once the captain sights the exotic beach of Alang, he just needs to look for a gap between all the other wrecks, line his hulk up and steam full ahead up the beach. He dumps his dirty load, flies home without having to think much about death. But death there is.

Over the last 23 years there have been 485 deaths in India, 460 in Bangladesh and 345 in Pakistan, according to Shipbreaking Platform, an NGO. These figures exclude all workers who died from illness from the hazardous cargoes abandoned in ships. It includes only the big reported accidents, such as when an oil tanker exploded on a beach in Pakistan on November 1, 2016, killing 77.

Deaths are inevitable when you cut oil tankers to pieces with hobby tools. On a beach, it is impossible to work with large machines and cranes. So in Bangladesh they send an army of workers, among whom disadvantaged children, into the ships armed with iron saws and hammers. The reason why the ships are broken down to pieces is that money can be made from recycling the steel. A large container ship fetches up to €10 million in India. And for steel there is no such thing as a consumer label. Nowadays, a chocolate brand has to think twice before going to market without some fair or green (washed) label on it. But even upstanding, clean consumer types like me do not care where the steel in their bike comes from. It goes to show that there are limits to the idea of making consumers more conscious about the products they buy. I too get tired of checking everything.

In most cases, it is just a whole lot more efficient and effective when states make arrangements to block dodgy products from the market in the first place.

Almost 40 percent of all ships in the world visit the EU, but only 1 percent are scrapped there. More than half of all major ocean ships in the world end up beached in Pakistan, India or Bangladesh. The EU estimates that each year, more than a million tonnes of toxic waste travels with those ships from the EU to South Asia. That includes 3000 tonnes of asbestos. In many yards, workers just burn it on the beach.

Europe has a couple of decent decommissioning yards. But the owners of knackered ships get far less if they retire them there. In Europe you get around $150 per tonne of steel, compared to $300 in China and $500 in India. Despite this, more and more Scandinavian companies are opting for the less lucrative but more sustainable option, under pressure from activists and journalists doing some decent reporting on the conditions at shipyards in South Asia. In addition to moral, ecological or social motives, there are also economic motives to dismantle ships closer to home.

At Europe's largest ship-recycling site in Ghent, around 35,000 tonnes of steel is recycled each year. You would expect the Belgian government to support European regulations prohibiting the dumping of ships on Indian beaches, in order to boost jobs at home. But Belgium actually favors weak regulation on the beaching of ships.

Perhaps that has something to do with some of Belgium's great and good. Take the Mediterranean Shipping Company (MSC), whose largest base is the port of Antwerp, Belgium. The company is based in Switzerland, a pioneer in secret bank accounts and amoral flows of cash. MSC is the world record holder in dumping toxic ships in developing countries; 55 to date. A fire on one of them in 2009 killed six.

The company came under fire from pressure group

Greenpeace 10 years ago. A Swiss TV documentary in 2012 confirmed that MSC was still playing dirty by sending ships containing asbestos and other hazardous materials to India. According to the University of St Gallen, it continues to violate numerous international conventions.[166] Yet MSC has been trying to cultivate an environmentally friendly image.

Attempts to try and force companies like MSC to get shipshape have been fruitless. The catchily named "International Basel Convention on the Control of Transboundary Movements of Hazardous Wastes and their Disposal" as well as the European Waste Shipment Regulation were quite literally circumvented. Shipping companies sold their ships as soon as they were outside European waters. The EU then banned ship dumping in India with its EU Regulation on Ship Recycling. But the back door has remained open since the rules only apply to ships that sail under the flag of a European nation. More than 90 percent of MSC ships are flagged to Panama, a modern-day pirate flag that quickly tells you that this ship is fishy. Panama's shipping regulations come with minimal taxes, environmental and working conditions. Having a Panama flag, or one for Liberia or the Marshall Islands, is the same as signaling ownership of a shadowy Swiss or offshore bank account. Today, 40 percent of all goods transported by ship are stowed aboard a vessel with one of these flags-of-convenience.

Tracking this cozy little scandal has for many years fallen to a little organization called the Shipbreaking Platform. Led by Patrizia Heidegger, a tireless woman working in an attic room just above my former office, the NGO has occupied an equally obscure corner of the environmental movement. It should not be this way. After all, shipping is to the industrialized economy what blood is to the body: it transports a huge proportion of the goods we all use every day. Both systems are vulnerable and choking a main artery can have major consequences.

Patrizia:

The shipping industry is regulated by the UN International Maritime Organization (IMO), but not well. New conventions always reflect the extremely low standards that Panama, Liberia and Marshall Islands want, whose combined fleet is so large that no convention reaches quorum without them. The IMO may be the only UN institution that regulates a single industrial sector, but in practice, industry is regulating itself.

In fact, the US is puppet master, controlling how these three weak states act. The US has historic and contemporary ties to all three of them that puts it in a position of very strong influence.

The IMO could become a powerful control body. In rare cases, it already is. Following a series of oil tanker disasters, it mandated structural improvements to this type of vessel. Even then, it was the EU that got the ball rolling after the Erika disaster in 1999 and it was the US that was impatient to ban the less robust type of tanker. The IMO had no choice but to take action. Enforcement was simple. Check compliance at port and if they were not ship shape, they were not coming in or out.

There is no way beaching can be done safely. According to the International Labour Organization, breaking up ships is the most dangerous occupation in the world, mainly due to the many deaths on beaches in South Asia. Aside from the many accidents, workers there are occasionally also confronted by the police, who have fired their guns on them when a rare demonstration ran out of control.

Banning beaching would be a great step forward and the good news is that in a technical sense, the solution is at hand. "You only need two things," Patrizia explained.

First, a set of technical standards for how to dismantle a ship. The EU actually does this with the Ship Recycling Regulation, which bans breaking up vessels on beaches, imposes strict rules on labor rights and on hazardous waste management.

But secondly, you need a license that obliges shipowners to deposit a certain amount on a blocked bank account every year, funds set aside for the vessel's eventual retirement. Each subsequent owner pays the same sum into the same blocked account and the last owner gets that money only if the ship goes to a proper decommissioning site, one that meets safety and environmental standards. If port authorities control compliance, this system could be in place and working as quickly as the tanker rules. To achieve all this, there just has to be political will.

In some places, the political will is growing. The world's largest sovereign wealth fund (i.e. Norwegian Central Bank – Government Pension Fund Global) decided to exclude companies that send their ships to be broken on tidal flats from their portfolio. With so many banks following the decisions taken by the Norwegians, that sets an important precedent.

In the shipping industry there's little will to change course. Cronyism reigns supreme. The shipping industry has pretty long tentacles. Take the Compagnie Maritime Belge (CMB). Like Umicore, the company got rich on the back of a genocide in Congo. King Leopold, who decimated the population of Congo from around 20 to 8 million during his brutal reign, was the one who deployed the then Compagnie Belge Maritime du Congo to ship the loot back to Belgium. As owners of CMB, the Saverys family inherited the massive wealth made on the back of this massive plundering and man slaughtering. As billionaires they are one of the wealthiest families in Belgium.

The Saverys's tentacles reach to De Tijd (the Belgian equivalent of the Financial Times) and the influential think tank Itinera, to name just a few branches of their business empire. So it is not in De Tijd that you will read how CMS's ship, the Mineral Water, switched ownerships and flags using a mailbox company in the British Virgin Islands (owned by shady Indian diamond tycoons, as shown in the Panama Papers) and the Polynesian Island of Niue to become the property of the company Kabir Steel in Bangladesh.[167] Before applying for a job at CMB I recommend reading 1984 by George Orwell. War is peace. Ignorance is strength. A toxic tanker is Mineral Water.

An investigation into possible criminal offenses committed by CMB in relation to "The Mineral Water" was opened. The CEO of CMB should start worrying when reading about the case of Seatrade in nearby Rotterdam. On March 15, 2018, the Rotterdam District Court sentenced, on the basis of the EU Waste Shipment Regulation, shipping company Seatrade for the illegal export of vessels sent for scrapping on the beaches of South Asia. The heavy fine and 1-year ban to execute a job in the sector was the first time that a European shipping company had been held criminally liable for having sold vessels for scrap to substandard shipbreaking yards in India and Bangladesh. More worrying for the CEO might be that the Prosecutor's request that the Seatrade executives face prison was only waived in light of this being the first time such criminal charges had been pressed. This groundbreaking judgment sets a European-wide precedent for holding ship owners accountable for knowingly selling vessels, via shady cash-buyers, for dirty and dangerous breaking in order to maximize profits.

* * *

Belgians are recycling champions. But the Flemish go a step further, recycling more than 65 percent of household waste. We are the global recycling elite. Many of us actually enjoy it, making a sport of how long we can do before putting a mixed waste bag on the sideway and for those that do not, there are strict rules to ensure we do our duty. Meanwhile, the European Commission says that a quarter of all waste exports from European ports to Africa and Asia are illegal exports. A large chunk of that leaves from Antwerp, Flanders' gateway to the world. Until Belgium manages to issue the first ever criminal penalty for dirty, dangerous and illegal dumping of ships, it seems harder to throw a beer can in the wrong place than to dump an oil tanker on a Bengal beach.

Chapter 8

How we give back to earth

In this second part of the book, our journey shifted to the end of our pipeline economy. We learned that we are fooling ourselves by trading hot air, creating poisonous circles and trying to bend a thick pipeline into a circle. This can only end in tears, scientists say, unless we drastically change course. Products with planned aging should simply be banned from the market. Illegal toxic waste streams to exotic locations should be a ticket to jail rather than riches. Laws to prevent all this should not be as leaky as Fukushima's nuclear power plant. It is foolish to wait for philanthropic business executives to improve our lot. The State will somehow have to save Mother Earth and I am afraid this will require some old-school legal punishments for those who are busy raping her.

If you look at the global economy from a human perspective, a perspective that envisages a thriving civilization at the end of this century, how might today's obese pipeline be transformed? Just try to imagine that for one reason or another, on some blue Monday morning, the most polluting half of all ships representing half of all world trade on earth are suddenly stuck in harbor. Imagine that large groups of citizens, supported by scientists, shift their focus from the start or the end of the global pipeline and instead point their arrows to the few choke points in between them? Any chain is only as strong as its weakest link and in the case of the global economy, the choke points are certain sea lanes and a limited number of major harbors. If the world economy is an hourglass, harbors are the narrow middle separating extraction sites at one end from shopping malls at the other.

In 2016, some 4000 people stopped a major coal mine in

Germany from working for a full 2 days. How many people are needed to block major ports? It took just 500 to block the biggest coal depot of Antwerp harbor for a full day, back in 2009. At the time it was the biggest action of the climate action camp in Belgium, in collaboration with groups from the Netherlands.

We also need networks to inspire conscious consumption, low impact idols on TV and more community supported farmers like Tom. They show us how to consume differently and provide solutions to disconnect us from the global pipeline that seems bound to blow sooner or later. But they alone are just not enough. As long as they merely chip away at the problem while the global pipeline remains vast and gaping, we are still on course for a disaster, all of us.

In the global struggle to reduce the volume of mining, consumption and waste, we will need to think strategically. Geography matters in every conflict and this global conflict is no exception. While exploitation and consumption occur almost everywhere, 90 percent of world trade depends on a limited number of ships, ports and routes. Most often, these choke points are located where a lot of people live. People that could be moved into direct action. We already have some experience in blocking boats, for example through Greenpeace and Sea Shepherd. Robin Hood style pirates could also disrupt shipping lanes in Europe with just a limited number of well-trained activists backed up by a landlubber legal team. Small but motivated groups of canoeists could form a series of clots in the world's veins and seriously constrict the flow. Precedents have been set from the US to Australia, sometimes with just two people blocking a coal port for a whole day.

Now imagine that you also have the unions on your side. They will need assurances over their jobs and that is why a detailed alternative employment plan should be worked out. Putting a brake on the excesses of global trade could actually stimulate untapped potential in regional trade and thus create more jobs, for example in the inland shipping sector. About 15 years ago I had a professor in urban planning called George Allaert. I recall how he thundered about the scandalous under-utilization of all the inland waterways from Paris to north of the Netherlands and into Germany. He stormed out of meetings because officials refused to fully harness the network that already exists. Inland shipping in Belgium only started improving in 2016 after a mileage tax was slapped for all cargo traffic by truck.[168] But the margin for growth remains enormous, according to experts such as Allaert.

If the above idea sounds outlandish, that may be in part due to a truly fundamental belief that all international trade is good *by* nature, regardless of what it means *for* nature. And that brings me to the third part of this book, where I will talk about the myths that are so pervasive that ideas like blocking ports hardly come up even in the minds of many environmentalists. These myths make the idea of slowing down global trade look stupid, terrifying or worse. I am going to talk about the basic assumptions that determine almost all our policies today and that together form the dominant world view or ideology of our times. In my humble opinion, this world view is operating in contradiction to physical reality, as described in parts 1 and 2. I am going to talk about world trade, major agreements to boost it, its ideological underpinnings and finally, the greatest myth of all, the perceived necessity for perpetual economic expansion in the form of GDP growth. What follows are the myths of this age.

Chapter 1

Myth 1: More world trade makes us stronger

Imagine you are walking across the drowned land of Saeftinghe, now part of the breathtaking banks of the Scheldt river, at the border between Belgium and the Netherlands. These brackish marshes are a breeding area for a variety of beautiful birds. The wind blows salty air through your hair and the high helm grass tickles your knees. The scene easily brings you to the edge of a trance until, suddenly, a container ship looms large. Three football fields long and as high as a suburban housing block, it is apparently sliding slowly through the tidal muds. In fact the Scheldt runs somewhere in between this vast expanse of salt-loving grasses and a maze of gullies. The mass of containers might as well be on an airport luggage belt, one that is in this case heading towards Antwerp. Today, the port of Antwerp receives over 200 million tonnes of goods a year. The next batch of stuff Made in China is always on the horizon. 24/7.

Once upon a time, things were very different here. Forget about the ancient Silk Road bringing in a few goods once every now and then. Very little stuff that was made in China ever made it to Belgium before World War II. Even the supply from colonies in Africa was a mere trifle compared to the floating hypermarkets that now arrive by boat every hour, day and night. After World War II, international trade started to grow exponentially. Between 1950 and 1992, the number of tonnes shipped over sea multiplied by a factor of eight.[169] On global trade, humanity entered a whole other level in those decades. But the time lag between trade's benefits and bills created a myth: the myth that more global trade always equals progress.

Fair enough: more free trade created great prosperity,

especially at the top. But after World War II economists were misled by temporarily cheap oil prices that made them believe that geographical limits were now a thing of the past. What we're witnessing now is revenge. The revenge of geography, the revenge of the climate and the revenge of the victims of unequal trading.

* * *

In 1804, the British sunk the Spanish frigate *Mercedes*, along with 500,000 gold and silver coins that were on board. Two hundred years later, they were found by US treasure hunters. A court ruled that the US treasure hunters had to hand over their $500 million booty to Spain. But just before the coins were about to be flown from Florida to Spain, the US Supreme Court received a novel claim from Peru. The country had been occupied and plundered at the direction of the Spanish crown and that armed robbery was neither forgotten nor forgiven. But the US Supreme Court ruled in favor of the conquistadores, thus renewing the legitimacy of the plundering that colonization was. Collective memories of the robbed have a very real meaning today – even if 2 centuries have passed.

Ships have a massive impact on our history and not always and only for the better. Without ships, the Spanish elite would not have been able to plunder Latin America, millions of slaves would not have been abducted from Africa and shipped to the Americas and we would not today be emptying Senegalese waters of its fish thanks to European super trawlers the size of small aircraft carriers. Australian sand would not be going into concrete for the world's highest tower in Dubai, and it would not be dumping our e-waste in Ghana and China. Many unfair and damaging trade flows, fed on over-extraction and encouraging overconsumption, are powered by just a few thousand mighty ships, who depend on just about a dozen major shipping hubs or knots in the global shipping lanes.

All along the way, the argument has been used that big ships are more efficient. But it's just like our refrigerators and cars that have become more energy efficient but also a lot bigger and heavier, thus undoing any benefit of the efficiency gain. More importantly: the so-called gains have allowed for new uses and new wastes. Efficiency is only useful in an ecological sense if it helps to get us to use less of our finite resources. Super-efficient shipping has done just the opposite. There are just not enough materials on earth to provide every earthling with the most efficient Tesla and iPhone.

Professor Alf Hornborg of Lund University, Sweden calculated that Europe has been a net importer of natural resources for centuries. His research also shows that Europe not only managed to maintain this position after decolonization, but expanded it massively. Europe is the most import dependent of all continents, followed by the US.[170] If you see how unequal trade is sucking life out of non-European countries, you could even argue that there is empirical evidence to the claim that Europe is the world's largest bloodsucker, unlike anything the world has ever seen.

Our dependency on materials, energy, food and so much more from other continents makes us subsidize dirty international shipping in exchange for the possibility of unsustainably high consumption, which is enjoyed mostly by a few at the top at the cost of billions elsewhere. But a majority of Europeans consider access to cheap exotic products a basic necessity and vote for political leaders who promise more of the same. With the problems exported beyond Europe, there's no democratic way in Europe to correct this unfair system – unless you count on a majority voting purely on moral grounds. Clearly, that is not happening. Neither do I expect that to happen.

Europeans amount to less than 10 percent of the world population. If you look at the voting turnout, it is more like 3 percent of the world population that somehow voted for this massive transfer of goods into Europe. Our economy has been globalized but democratic constituencies remained regional, and flawed.

Let us spend some more time on the consequences of the current trade volumes. One consequence of all this freedom and blank checks for shipping is that the 16 largest ships in the world emit more sulfur than all cars together.[171] Yes, all billion cars. International shipping is not only tax-free, it is also free to poison our atmosphere. Ships are allowed to use unfiltered bunker fuel that is literally thousands of times more filthy for the air than what is used in the average car. James Corbett of Delaware University, a world authority on ship emissions, estimates the worldwide death toll from ship emissions at 60,000 victims a year, of whom 27,000 are Europeans. The idea that these ships can pollute the air above the ocean without having an effect on humans is fatally flawed and the flaw has a name: wind. Ships don't stay in the ocean forever, they even usually come to shores right where many people live. Shipping emissions have continued to rise since Corbett published his groundbreaking paper in 2007 and we are probably now looking at almost 100 European deaths *every single day*. From shipping.

It is a similar story with CO_2 emissions. Almost every sector and industry has been required to reduce emissions (or pay for fake carbon offsets). Not so for international shipping. As a result, studies show emissions from shipping will increase fourfold between 1990 and 2050.

The experts know all this. But at the 2015 climate summit in Paris, shipping and air traffic were again sitting pretty, exempted from action to reduce their emissions. A package of measures to further reduce emissions of road vehicles was adopted by the European Commission in 2016. That too was a missed opportunity to lock horns with the aviation and shipping sectors. The UN Maritime Administration decided at the end of

2016 to postpone emissions reductions from shipping for at least 7 years. Climate urgency? What urgency?

Why would the powers that be give shipping such a free ride? Could it be because they are at the very heart of the macroeconomic model in which each product has to be made where the cost of extracting or making the product is at its lowest? The invisible hand must be free to show us its magic tricks. Just to be clear, I do believe that some world trade is necessary for our well-being. I could not go without coffee, to name just one daily indulgence. But the question needs to be asked: what is the optimum trade volume? Is it a volume that allows for a level of specialization and prosperity, but also makes over-extraction and overconsumption hard instead of easy? A volume that allows our ecosystems to regenerate so we can carry on enjoying their benefits forever?

Why not take as a benchmark Earth Overshoot Day, the day in the calendar year after which we are eating into our capital, as the rent for that year has been used up by then. On that measure, one could argue that the optimal volume of the global economy was reached in around 1970. In 1987 we were still close to balance but from there onwards the speed of taking more from the earth than it can regenerate has been exponential. Now, it's important to understand the scale of the problem. The exports of goods didn't just double or triple ever since. In dollar value, world exports in 2016 were 50 times bigger than in 1970.[172]

Returning to 1970s trade volumes does not mean trading the same Ford Cortinas, flared jeans, disco vinyl and other 70s stuff. Once a manageable volume is agreed upon, many questions can be asked around quality, fairness and ecological sense. A Ugandan farmer making money from fair trade bananas sounds to me like a better idea than building a greenhouse in my garden that is heated to 30°C throughout the year. It also makes sense to keep trading in bananas, given their contribution to a healthy diet. But should our ships really be made in China, used in Europe and broken down in Bangladesh? Should shrimp caught in the North Sea get trucked to Morocco to be peeled then trucked back again? In that process it is not just the climate that suffers. We lose jobs and gain traffic jams, excessive packaging, emissions and the associated avoidable air pollution diseases and deaths in return. It adds GDP, but costs the public more than it benefits the shareholders of the companies who do just that. Any policy that is serious about both climate change and bringing back jobs must say the unthinkable; that we need the right mix between world trade and protectionism.

Protectionism? Relax. Breathe in...and out. I know that some readers may get a little hot under the collar reading this. Protectionism is, after all, embodied by fascists like Donald Trump and Steve Bannon. That should not be the case. Influential Belgian Green Party member and columnist Jan Mertens wrote a strong piece about the misplaced fear of green progressives who were anxious not to be associated with the taboo word, protectionism.[173] "You sure need to question policies based on 'own people first'...but at the same time, that does not mean that you cannot talk about the deeper meaning and consequences of social and environmental exchanges in the form of international trade. Faith, because that's what it essentially is, in the blessings of unfettered international free trade is dangerous. We need to have that debate without being lumped together with the likes of Donald Trump."

Protectionism is not the same as patriotism. Patriotism is about nationalism but protectionism can be about protecting the global ecosystem from a collapse. According to Nobel Prize winning economist Joseph Stiglitz, you can apply existing WTO procedures for tariff walls around polluting countries. Instead of charging an import tax on solar panels from China, why not penalize damaging imports like bauxite, uranium, gold, oil, gas, coal, wood pellets and palm oil? But things also need to go further than taxing. Why should the international transport of pellets even be legal? Why should gadgets with hidden kill switches be allowed on the market?

Will all this not hurt the poorest part of the population hardest? Not necessarily. That depends on how you spend the extra import taxes and how you organize your economy in general. Guaranteeing a good living standard for all and a limit on luxury goods merely requires political imagination. No laws of nature need be broken. It is if we fail to intervene that systems will collapse, with super inequality as a result and the poorest starving to death. A much fairer distribution of the raw materials, through a global trade regime that respects the capacities of our ecosystems, can deliver more people more prosperity over a longer period than today's free trade mantra. Governments need to step up and co-operate to create ecological trade barriers, instead of negotiating on what trade barriers they will eliminate first.

* * *

What does all this say about the so-called free trade agreements that most Western leaders are so keen about? The two big cross-Atlantic ones are known best by their acronyms: TTIP and CETA, standing for Transatlantic Trade and Investment Partnership and Comprehensive Economic and Trade Agreement respectively. They are two of the approximately 30 agreements that the EU wants to conclude. All want to further increase the flow of goods to and from the EU.

Most of the classically big conservative, liberal and even many social parties on the European continent still believe in the myth that TTIP and CETA are good for us. They discuss the details of what trade barriers to eliminate but not the fundamental problem with current trade volumes.

The threat from TTIP seemed less acute when Trump took office and immediately poured cold water over it. But the threats are anything but over. First of all: CETA is TTIP through the back door. About 42,000 US multinationals have a branch in Canada. If CETA is pushed through with an arbitration panel, those companies would also be entitled to claim damages of up to several billion euros against any new EU law that curbs profits of US companies, however ill gotten. The watchdog NGO Corporate Europe Observatory, in its report *The Great CETA Swindle*, denounced the Canadian government and the European Commission's efforts to sell the deal as a "progressive" agreement. CETA remains what it has always been: an attack on democracy, employees and the environment.[174] A total of 101 law professors warn that it will result in a transfer of billions from states to multinationals.[175] For more on the spine-chilling power grab of secretive trade deal courts, look no further than The Court that Rules the World, a BuzzFeed report that was 18 months in the making.[176]

Belgian professor Paul De Grauwe of the London School of Economics thinks all so-called free trade agreements should be put on ice for a number of reasons, but especially because of their environmental impacts.[177] Another Belgian professor, Jonathan Holslag, wrote of CETA: "We forget...all the hidden costs such as the inefficiency of the endlessly long distribution networks and the associated mountain of packaging waste and energy-loss for cold storage transport."[178]

* * *

One of the key features of CETA is that it opens all areas of health, environment, food safety and social protection up for privatization, unless something is explicitly put on a safe list of regulations that is set in stone at the start. So the European Commission now calls all consumer protection and environmental laws non-tariff barriers to trade. What protects us and the earth is a "barrier", de facto a "bad thing". A ban on the carcinogenic weed killer Round-Up? Bad. A ban on tar sand oils? Bad. Trade is good. Barriers are bad. And all the sheep from Orwell's Animal Farm repeat that mantra all over the place until we're totally brainwashed. Trade is good. Barriers are bad.

Just negotiating TTIP and CETA has already had massive impacts. The European Commission approved a whole series of genetically modified crops (GMCs) for sale in the EU in 2015 as a concession to US negotiators and a step towards harmonization. A better choice of words would be: as part of a race to the bottom. The European Commission also dropped regulations that would have side-lined dirty tar sand oil, to please Canadian negotiators. It's all in leaked official documents.[179]

Also on the chopping block: the precautionary principle, a tenet that policymakers can ban a particular product if there is a high risk of damage, even if there is no existing scientific proof of the damaging effect yet. Currently, the burden of proof lies with the company to prove that its goods are really safe. CETA and TTIP are based on the reverse thinking, which is customary in the US and Canada. Harm must first be proved by the state before a policymaker can ban a product.

More than 3 million Europeans signed a petition against TTIP and CETA. Protests have been among the largest seen on the continent in recent years. Cecilia Malmström, European Commissioner for Trade, reacted with this: "I do not get my mandate from European voters." If the European Commission ever needed an epitaph for its gravestone, it need look no further.

* * *

Malmström had a point: she's not elected. But she does hold a lot of power. How come? Problem one: the vast majority of eligible Europeans did not bother to vote for a member of the European Parliament last time round. The 2014 elections reached a record low, at just 42.5 percent. More than a third of those who made the effort voted for an anti-European party. Only a quarter of eligible voters took the effort to go and vote for a pro-EU candidate.

After that is all done, the strongmen of the biggest pro-Europe parties decided who they want in the club that we call the European Commission. The European Parliament has close to zero impact on that. But only the commission can draft new laws and its president is the de facto prime minister of Europe. The checks on this system today are not a strong parliament, but a couple of anti-EU nationalist heads of state who try to sabotage the European Commission in the European Council – often on the already weak environmental initiatives they do propose to take.

So if Cecilia Malmström does not work for European voters, who does she work for? It is hard to keep up with the number of commissioners who, at the end of their term in office, score an extremely well-paid role in a large company with plush offices. Just take the previous boss, José Manuel Barroso. He is now working for Goldman Sachs, a bank that, according to the commission's current boss, was partly to blame for the recent financial meltdown. Moreover, it played a very dubious role in cooking the books in order to allow Greece to join the EU in the first place, a fiddle that would come back to haunt Europe. Leaked emails show that during his mandate, Barroso met people from Goldman Sachs much more than official documents suggested. As Peter Mertens, leader of Belgium's fast rising left-wing party PVDA, writes in Graailand: "The business bank that has been responsible for the major banking crisis has given advice to the President of the European Commission on how to address this banking crisis." Barroso is no exception. Former European Commissioner for Trade Charles De Gucht went on to receive €144,000 a year from ArcelorMittal, a steel company that has a major interest in not paying the real total costs that come with shipping stuff around the globe. CETA and TTIP are good for our wallets, say people like Karel De Gucht. He probably had his wallet in mind.

* * *

The arrogance of the European political ruling elite, and even of the majority in the European Parliament who voted in favor of CETA, is a total disaster for the EU as a project. It is a disaster because this arrogance has brought Europeans to the point where they are ready to throw the baby out with the bathwater by dumping the European project wholesale. If the entire European project comes to a dramatic end, there will of course be other reasons besides rotten trade deals. But TTIP, CETA and the lack of democratic legitimacy and respect for the democratic expression of Europeans has alienated millions of Europeans towards what was once considered a beacon of progressive governance.

In October 2016, Wallonia made its big stand against CETA, standing alone in Europe. The pressure was shouldered by the then heroic minister-president of Wallonia: Paul Magnette. The great and the good of Europe's political aristocracy put in calls to try and railroad Magnette into backing down. The President of the Council of Europe, Donald Tusk, issued several impossible and undemocratic deadlines to Wallonia. Yet, little Wallonia (and the Brussels region as well) stood firm and forced a number of concessions. It's thanks to them that at least the ECJ is now looking into the basic question of whether CETA is legal or not.

* * *

Tatiana Santos works for the EEB on European chemicals policy, lobbying on behalf of 30 million environmentally conscious Europeans.

Scandinavian countries had to water down their chemical standards when Europe adopted new rules called REACH. Although this raised the bar for many countries with weak regulations, that is only useful if the laws are put into effect. But in recent years, the banning or phasing out of hazardous chemicals has ground to a halt. Two forces are driving this; lobbying by a strong chemical industry, and TTIP. The result is that hormone disrupting chemicals are still making the environment and people sick.

The report *A toxic affair* describes this in detail.[180] Hormone disrupting chemicals are causing diseases, deaths and medical costs to the tune of over €150 billion every year in the EU. By comparison, of 60,000 known industrial chemicals, only five are forbidden in the US. In the EU, thousands are either regulated or forbidden. No wonder that REACH legislation has attracted a record number of "trade barrier" procedures by the US. Just negotiating TTIP is hacking away even at the already limited regulations.

Similar pressure is being applied to European fiscal policies. The Robin Hood Tax – a tax on financial transactions long in the making that would provide financial stability and money for the economy outside the finance bubble – was put on hold due to the trade deal negotiations. The Robin Hood Tax would take money from the rich speculative class of people who want to make even more money with their money without having to actually work for it. At one point it seemed that a long struggle to get such a tax was finally coming to fruition in Belgium and ten other EU countries. But one bizarre element in this story is that worldwide one of the great proponents of this Tobin tax is the same man who suddenly started obstructing: Belgian Prime Minister Charles Michel. In 2010, Michel fought within the UN for a global financial transaction tax that would bring over €30 billion to fight "injustices". He was still in charge of development affairs then, so it would have been great for his portfolio. But even back in 2004, Belgium was the first in the world to introduce such a tax – even if the tax would not come into effect until all other European countries did the same. When that seemed too much to be asked, there was a European initiative to start with a group of countries. This tax would make Belgium more than €1.36 billion richer. However, a leaked document on the TTIP negotiations talks about reducing obstacles to the financial sector. According to the Corporate Europe Observatory, it aims at the financial transaction tax even before it is there.[181]

Trump took a hammer to US trade agreements with Pacific countries, with Canada and Mexico and with Europe. He then started a trade war and at first it seemed that the flow of goods across the oceans would not continue to rise as fast as it did under the watch of his predecessors. But with Trump there's no strategy for reducing ecological impacts, rather to the contrary. So it shouldn't have come as a surprise when he suddenly brokered a deal with Jean-Claude Juncker in the summer of 2018 to not only make trade war exceptions for the EU but even talked about zero tariffs and zero subsidies as the ultimate common aim. These gentlemen are now deciding that rather than going for a pretense of democratic control by burdensome parliaments, they'll just deal with trade issues in closed door meetings. One thing they decided is that the EU will buy more liquid gas from the US, thus allowing the US fracking industry to wreak even more havoc.

* * *

In short, trade agreements aiming to increase the trade flows across the oceans even further are hitting the environment on a global scale, reducing environmental protections, eroding democracy and support for the European project, increasing inequality, deteriorating our food safety, attacking our health through the use of more hazardous chemicals, undermining financial stability, privatizing anything not on a "safe list" and leading to the confiscation of billions of taxpayer funds by multinationals that are allowed to play victims of policies that try to protect people and the environment. These are all not some vague opinions or leftist ideas but facts.

The arguments in favor of a deal like CETA? Europeans and Canadians will recognize each other's professional qualifications and emigrate more easily. People migrating more easily is a liberal idea and I'm happy to see that happen, but most of the features of TTIP and CETA come right from the neoliberal cookbook. And that brings us to the second great myth of our times, one that is crucial to understand if we ever want to be serious about the challenges described in this book.

Chapter 2

Myth 2: Neoliberalism is the way to freedom

At the end of 2014, a strange article appeared in The Economist: *Ships that pass in the night*. As it was about TTIP, I was curious if it mentioned any of the Titanic drawbacks of long-distance transport we have covered. But no, The Economist was concerned with insufficient slime trade. San Francisco oysters, it claimed, are every bit as delicious as those of Normandy. But those poor French are being wrongly deprived of them and TTIP will finally do something about this problem.

I found so much wrong in that story that I've started using it in trade debates over and over again. Only a journalist that is unaware of the fact that the metabolism of our economy is already way too high can come up with this shit. "The consumer is king" narrative is killing us all and shipping oysters from San Fran to Paris in cooled containers is exactly the kind of trade we just can't afford ecologically. Such articles, as well as TTIP and CETA, are the children of neoliberalism.

But what is neoliberalism? Essentially, this ideology states that a government must function like a company. According to Wikipedia, neoliberalism is a collection of ideas about how to manage the world economy. Four principles stand out: privatization, minimal government spending, deregulation and so-called free trade. These assumptions, axioms or dogmas are the four commandments of the neoliberal ideology, a toolkit to engineer unlimited economic expansion. It boils down to putting capitalism on steroids.

For neoliberals, all economic discussion can be reduced to the speed, order and details of implementation of those four commandments. They claim this is the road to freedom. Free is their sacred verb: free markets and free people. So dominant is this ideology that its adherents do not even call it an ideology at all. To do so would acknowledge that there are competing alternatives to steer the economy. There's a long line of politicians from Margaret Thatcher in the UK to Bart De Wever in Belgium who try to kill economic debates by crying TINA: There Is No Alternative.

That's the myth, but what's the reality? Money must also be free to go where it wants to go. We know that through the various scandals covered by the press, such as Offshore Leaks, LuxLeaks, SwissLeaks and the Panama and Paradise Papers. Belgium has lost out on billions of tax income through the international financial architecture that allows the rich to park their money in tax havens. It is estimated that globally, tax havens have allowed a transfer of €18 trillion. That is money taken away from the great leveler: taxation. It is people with a big capital making an even bigger capital. Trillion is a big number, not used often and it has different meanings in the US and UK so just to be clear, we are talking about €18,000,000,000,000. This is a lot of money that states couldn't tax. States that, according to this class of tax evaders, need to solve their financial problems by spending less on frivolous things like healthcare, public transport, pensions and so on. To justify the dismantling of the welfare state various names are used: austerity, axing, solving the structural deficit etc. Neoliberals are good at victim blaming: if the budget has a hole they'll blame those profiteering ill and pensioned people.

Under neoliberalism, private profits trump public losses. A good example is Dieselgate. Europe has emissions standards for cars to limit things like fine dust to reduce the risk of getting lung cancer and other diseases. That standard is supposed to shield us from the 400,000 dirty air deaths that Europe endures every year, of which 11,000 are in Belgium.[182] In this small country, more than one person dies every hour because of air pollution. But there is more to diesel than how it kills legions of citizens in a rather direct way. Diesel cars are also the main source of black carbon, which causes the North Pole to warm up much faster than the rest of the world. Greenland was never really that green but today it should perhaps be called Greyland. Black carbon soot covers much of the ice, absorbing warm sunbeams and melting the ice faster than what can be explained by rising temperatures alone. The melting of the Greenlandic icecap is just one more piece of collateral damage of the private gains of car manufacturers. After Dieselgate, European policymakers undertook a balancing act. They then set new rules allowing car manufacturers to emit twice as much as before. If you decry this measure as legalization of mass murder, you get the swift response that the automotive industry, private interests, is important for our economy. Meanwhile, consumer, air pollution and climate fraud continue to grow. In 2001, the deviation between claimed emissions and reality was about 9 percent. By 2015, it was 42 percent.[183] Thanks to ever more creative ways to count the "normal use of a car". Climate progress over the last 4 years? Zip. A commission from the European Parliament investigated the Dieselgate scandal and the response of European policymakers. Its conclusion: "Totally inadequate".[184] The European Commission had been informed about the cheating by Volkswagen since 2012. It did nothing. Even after the scandal came to light, the response was inadequate. One wonders why Volkswagen models enjoy favorable exceptions even to the hopelessly generous new rules. The answer was found in a Microsoft Word document containing the favorable amendment. Its author: the Volkswagen Group.[185]

But it's not just cars. The European Commission is so keen on the neoliberal logic of putting private profits before public losses that it wants to submit all European laws to a test. This is dubbed the "better regulation" agenda – but it boils down to scrapping or calling down regulations that stand in the way of GDP growth. Never before has the European Commission showed such a neoliberal zeal than under the watch of Jean-

Claude Juncker. It is notable that the commission president is from Luxembourg. His deputy, Franz Timmermans, is from the Netherlands, as is the influential Jeroen Dijsselbloem, the former chairman of the EU finance ministers club. Luxembourg and the Netherlands are also the EU's premier tax havens. In a global ranking of tax havens, the Netherlands is ranked third in the world, just after Bermuda and the Cayman Islands.[186] The role of Jeroen Dijsselbloem in guiding the Netherlands to this position was briefly discussed in the first part of this book. At the beginning of 2017, documents appeared showing that Jean-Claude Juncker had for years been blocking all efforts to write new European rules to halt tax evasion.[187] This was during a period in which he held roles first as Minister of Finance of Luxembourg, then Prime Minister. The only other EU country that regularly supported his sabotage was the Netherlands, with Dijsselbloem at its financial helm. And these two were like the prime minister and finance minister of the EU for many years.

The situation is dire but simple. The European Commission under Jean-Claude Juncker is like a crew of fiscal pirates on a war path. Behind closed doors they are always on the lookout for emptying state coffers but in front of the cameras they complain about the state coffers being too empty. And when they find a state coffer too empty they'll tell that state to cut social welfare schemes. We, the 99.9 percent of Europeans without private bankers and tax havens, are chained up below decks, forced to row, no matter where the ship is headed to. And we have to row harder and longer with every new conquest of the pirates.

* * *

The growth of inequality within Western societies is not in question. It is statistically evident and acknowledged even by the neoliberal IMF as a major problem. Tax evasion is one driver of inequality, but so is privatization of basic services. But what are the precise consequences of greater inequality? In their book, The Spirit Level – why more equal societies almost always do better, professors Wilkinson and Pickett use mathematics to prove that a country with less financial inequality also enjoys less illnesses, social problems, mental health problems, drug use, obesity, academic underachievement, teenage pregnancies, murders and prisoners. With equality comes trust, a longer lifespan and greater compassion for the plight of other countries. The study does not say we should all have the same income, just that society is better off when inequality is relatively small. But the opposite is happening. Celebrated economist Thomas Piketty raised the alarm in his masterwork Capital in the 21st Century, which showed that societies will continue becoming more unequal for as long as neoliberal policies continue. There's no well-being for all within planetary limits as long as neoliberalism is our dominant ideology. There's only extreme well-being for a few.

But neoliberalism is not just about the regulation, economic choices, financial flows and inequality. It is also about a society in which a particular mental attitude prevails. We learned a lot about that from psychology professor Paul Verhaegen. His verdict:

Not so long ago, our culture and thus our identity were determined by an interaction between four spheres: politics, religion, economics and art, where politics and religion were fighting for power. Today, politicians are shredded by comedians, religion reminds us of suicide terrorism or sexual abuse and everyone is an artist. Only the economy is holy and the neoliberal economic story determines everything.[188]

More importantly, Verhaegen wrote about neoliberalism in The Guardian, outlining the perfect characteristics to help land a job in today's economy: talkative, narcissistic, an accomplished liar without remorse, flexible, impulsive and, above all, a risk taker. His punchline? These six characteristics are also the six key ways to recognize a psychopath. My 50 cents? These traits make me think of a certain Donald J. Trump. Trump is the kind of leader that neoliberalism leads to.

Under neoliberalism, there is only room for strong individuals. One of its most forceful proponents, former prime minister Margaret Thatcher, literally said: "There is no such thing as society." The idea that each individual only gets what they deserve has been elevated to a law of nature. There is no tolerance for bad luck. Set back by chronic fatigue syndrome, a car accident or being fired at 50 with precious little chance of finding a new employer? Tough luck. The ongoing epidemic of burn-outs is not recognized as the result of ever-increasing demands on workers. To the contrary, too often fingers are even pointed to the victims, accusing them of weakness.

Instead of society, neoliberals don't see a diversity of citizens, they only see consumers and entrepreneurs. Both of them need to perform, no matter what. Each individual must consume and may not go bankrupt, expectations that are the mirror image of a private company. Neoliberals are in fact counter-revolutionaries, not that they would care to admit it (they rarely even admit being a neoliberal). The revolution they seek to overturn is the French Revolution. Liberty, equality and fraternity: the holy trinity of this revolution is under attack from all sides. The French Revolution brought to an end the dark feudal era. If neoliberalism continues its ascent, history books may as well place us at the tail end of an all-too-brief era of enlightenment.

* * *

After decades of nurturing the neoliberal script, the IMF did a little introspection in 2016. Three of its researchers published a paper entitled "Neoliberalism: Oversold?" They concluded that the recipes of freedom of money to flow where its owners want it to flow, coupled with government austerity measures had together stoked inequality. The biggest surprise was how long it took for the penny to drop. But for the authors, this rising inequality is just unfortunate, not a serious problem. According to them, inequality is bad "because it stands in the way of further economic growth".

Two years later the IMF came with another confession, saying that, "The growing economic wealth and power of big companies—from airlines to pharmaceuticals to high-tech companies—has raised concerns about too much concentration and market power in the hands of too few." The authors claim that "some rethinking of policy is needed". And by the end of 2018 the IMF basically said that the UK was broke because it privatized too much.

These kinds of admissions are weapons in the hands of environmentalists who look at the big picture and the issue of power. Unfortunately, I've experienced that the n-word is taboo even in many environmental organizations. I have been laughed at for proposing to use it more often. Too politically loaded, goes the counterargument. It will push us into a radical corner, exclude us from serious political debate and policy influence. But neoliberalism is an economic theory that simply does not understand how nature works and as it is such a dominant theory, it is the source of a massive range of bad decisions with a negative impact on nature. Attaching a monetary value to things that just cannot be reduced to financial terms is a case in point. How do you calculate the value of an amoeba, an ant, the bees? Climate change is the biggest black hole in the neoliberalism universe. It's a market failure. What's the price of annihilating most life on earth, possibly including humans, during the twenty-first century? Of course, you can discuss the strategy on when and where to use the n-word itself and it will not help in many lobbying contexts, but I concluded you simply can't ignore the elephant in the room if you want to be serious about trying to steer humanity away from the current route towards the abyss.

Given the obvious failures of their theory, how have neoliberals been able to sell austerity, deregulation, privatization and more free trade with as an aim, more GDP growth? Neoliberals offer individual freedom in spades. Examples: the freedom to pollute and evade taxes. But they don't explain that what they offer is more like a trade-off between freedom and security. The influential ecological thinker Dirk Holemans writes in his book Freedom & Security in a complex world (Dutch: *Vrijheid en Zekerheid*): "On the way to more individual freedom, we have lost much of the security that led to modern civilization...With the neoliberal period, the nation state lost power and meaning, largely becoming a servant of the demands of multinationals."

At his death, US media and top politicians called Milton Friedman, the godfather of neoliberalism, the man of freedom. He's also the hero of Belgian Finance Minister Johan Van Overtveldt, who wrote a book about him. But the first leader who applied Friedman's recipe was General Pinochet, immediately after wresting control of the government from democratically elected leader Salvador Allende. The second great neoliberal thinker, Friedrich Hayek, schooled Pinochet directly. The shock doctrine of fast privatization combined with harsh austerity was combined with cruelty on a massive scale – to control the protests resulting from the coup and shock doctrine. Hayek came away from their meetings saying that he preferred a neoliberal dictator to a democratic government without neoliberalism. Since then, this troubled continent and the world at large have seen a great deal of shock therapy. So far for the "freedom" in neoliberalism.

The millions who read Naomi Klein's book The Shock Doctrine know that neoliberalism delivers the exact opposite of freedom. She gives examples ranging from juntas in South

America to Boris Yeltsin of Russia, who attacked the country's parliament with tanks. From the UK's Iron Lady to George Bush in Iraq and the US itself. Wherever they went, neoliberal leaders prescribed the same treatment: shock. Military, economic and sometimes even literally: electric shocks. The CIA started back in the 1950s to test the use of electrical shocks to torture and brainwash people. This enhanced interrogation technique (read: torture) was still in use in the Iraq war. Shocks were needed to rein in people's freedom and limit the power of public representatives. 9/11 opened the door to the Patriot Act, which allowed for the existence of Guantanamo Bay. Klein tears to pieces the myth that deregulated capitalism and freedom go hand in hand and that free markets complement democracy. She shows that fundamentalist capitalism – what neoliberalism boils down to – has consistently been implemented "through the most brutal forms of compulsion".

* * *

To put the ideas on world trade upside down one needs to understand which ideology stands in the way of doing that. To understand why neoliberalism has caught on since the 1970s, one needs to understand the even bigger picture of what it is supposed to deliver: economic growth. It would not be correct to blame all our problems on neoliberalism. The biggest company in the world, probably ten times more valuable than Apple and contributing to global warming like no other, is a state-owned company: the Saudi oil company Aramco. But the Exxons and the Aramcos of this world all have a greater and more mysterious god, one that seems to sit very comfortably on its throne. It is a god worshiped by both neoliberals, oil-oligarchs and communists alike and it's that god that is killing us like nothing else. The god I'm talking about is expressed as the GDP of today compared to the GDP of 3 months or one year ago. But to keep things simple, let us just give it the name that we've been hearing all over again, for decades on end: Growth (capital G intended).

Chapter 3

Myth 3: Economic growth is the key to progress

A 20 percent increase in the death rate of Americans over the age of 65 would cause our per capita growth rate to accelerate.[189]

This morbid quote comes from The Economist. Following this logic, it is also good for GDP growth if the government denies those aged over 65 access to hospitals and gives them a free cyanide pill. It would be a win-win situation: GDP increases through savings in social security and a booming cyanide industry would also add to GDP.

As if that is not troubling enough, the quote comes from an article on climate change. Basically, a magazine that avidly promotes GDP growth now writes that global warming will kill those that do not contribute to GDP but are a cost to society. In other words: progress.

Fans of Growth do not count deaths by the dozens, but by population percentages. They don't blink for a little bit of generational genocide. The folks I am talking about pray to Growth. The great creator and destroyer of wealth is now GDP and any pause in its growth is reason for played up mass hysteria and even more human sacrifices by the leaders who worship this god.

While neoliberalism is an ideology to radicalize capitalism, with institutes such as the World Bank and the IMF to enforce implementation, the growth religion is even bigger, vaguer and more elusive. It is the one god that unites leaders from Washington to Beijing.

It is easy to overlook that in historical perspective, that is a remarkable feat. After all, Growth is a newcomer. While God and Allah have been widespread and dominant for centuries or even millennia, the religion around Growth was just a peripheral phenomenon at the beginning of the industrial revolution. Two centuries later, Growth is the dominant religion in almost all countries on earth. On Sundays, people in Belgium used to go to church to be with God. Now they go to shops on Sunday to contribute to Growth. Consumption is sold to us as a mark of good citizenship. We must all do our bit for Growth. TV news reporters are grim faced when they announce a period of reduced consumer confidence. A drop on spending is like a bad weather forecast. Shopping is a civil duty.

Growth is fueled not only on ever more consumption, but also ever lower labor costs. As there is no bottom for this, it has been taken to extremes. In Germany, the drive to lower labor costs created millions of €1 per hour jobs. Few people realize that no other rich country has added so many working poor to its labor force. There are already 7.5 million people working in Germany for €400 a month. In 2016, the Germans went a step further, offering asylum seekers work for just €0.80 per hour. Millions of people are gently and less gently forced to work for (almost) nothing to fuel German GDP and exports. That should ring a bell with anyone who paid attention in history classes.

* * *

There have been times when a few establishment figures have questioned the neoliberal orthodoxy. Sicco Mansholt was one of the first European Commission presidents. In his position as the boss of Europe, he had a rather sudden change of mind, after reading the Meadows' report *Limits to Growth* to the Club of Rome. All of a sudden and to the astonishment of his cabinet, he started advocating against GDP growth as a policy objective. He proposed new barriers to imports of environmentally unfriendly products. He wanted to increase what TTIP wants to decrease: the flow of goods across oceans. Most people who know about Sicco Mansholt will know about what he did as the powerful agriculture commissioner: getting the EU from a food importer to a food exporter, in the process also creating the first and unfortunate butter mountains. But rather less known is what he did in his last and controversial year in the European Commission, when he was leading it. Frank Westerman's widely acclaimed book *The Republic of Grain* covers the transformation of Sicco into a statesman that acknowledged the science that told him that there are limits to growth and to trade.

The hegemony of the growth cult has obscured our historic view on this. Going back to Adam Smith, widely acclaimed to be the ideological founder of capitalism, you'll discover that even he did not think an economy could grow forever. He wrote that capitalism would be only a temporary phase en route to a sufficient-for-all economy. But how long is temporary?

* * *

The man who told me about what Sicco Mansholt did, Joan Martinez Alier, is a pioneer himself. Back in 1989, Alier was one of the founders of the International Society for Ecological Economics. Thousands of anti-neoliberal economists are now part of it. He taught me to look at the economy in a completely different way, one that is remarkably simple, yet transformative.

In its essence, Joan explained there are three "layers" in the economy. There is a hocus-pocus layer floating high above the real economy. This is the flashy, get-rich-quick world of the City of London and Wall Street, a shady realm of venture and vulture funds, credit default swaps, smoke and mirrors made to disguise the fact this is about using other peoples' money to make more money, regardless of the impact on the real world. Because this hocus-pocus economy is all about blowing bubbles and then profiting from them and then again from their bursts, this economy has a habit of crashing down to earth from time to time, on the heads of many victims in the real economy, in the real world. When the housing bubble burst in 2008, 6 million US citizens lost their homes – to give just one example.

Below that layer you have the so-called "real economy" of goods or services, brands and packaging, the products as in the literal meaning of the P in GDP. These are the things most people can name. But what most people don't realize is that in this economy, every oil leak or traffic accident is a plus. It doesn't look at capital and interest in an ecosystem sense. It doesn't internalize the real total external costs in the price of its products, who stand miraculously independent of the full costs of pollution they cause. Not that it can and must do that. No, the problem is rather that the wrong conclusions are taken from there. Rather than acknowledging that you can't put a price on killing people or reducing biodiversity and must thus take any measure needed to avoid an economic process that does that, economists operating in this so-called real economy will keep claiming it is only a matter of internalizing the cost in the product. Again: what's the cost of extinction then?

But the fundaments on which all of the above depend is what Joan called the real-real economy: the products and services provided by ecosystem earth. The earth produces and absorbs, gives and takes. The water cycle, the carbon cycle: it is the earth's processes that people make use of and depend on. The GDP economy is a bad guide to the state of this earth economy because it often gives positive values to matters that are negative for the real-real economy. Earth's economy has cycles (with also rapid changes) that we don't understand fully and can hardly predict. What we do know and witness is that there are limits to how far these cycles can be stretched, influenced, altered. Many of the earth's cycles are like strings: you can stretch them until they suddenly break. The task of the ecological economist is to study this earth economy and how and where humanity is stretching it beyond its limits.

This fascinates me. Joan teaches me how geography and economy come together or how nature laws – he often cites the laws of thermodynamics – can counter subjective opinions, wishes and delusions, such as the idea that indefinite GDP growth is possible on a planet with finite resources. As he says: "the industrial economy is not circular, it is entropic". I am familiar with that as I was originally trained as a physical geographer. In a world of warped ideologies and blatant lies, it gives me some peace of mind that there are still scientific certainties we can fall back on. We must use them to expose the lies and myths of our times. And then to build a new narrative.

On holiday in Barcelona, I leave my family at the beach to visit Joan at his home. The streets around his apartment are dotted with strange shops. Chinese stores sell clothes from the cardboard boxes they were shipped in. The overflowing neon lit garage-shops stand in stark contrast to the sight just above the stores: stately apartments with rounded balconies and beautiful wooden shutters. A century ago, this is where Barcelona's textile barons lived, when the city still made its own clothes.

Sharing this small observation with Joan was all he needed to get started on the ills of the world economy. "While trade flows need to go down to put the brakes on the sixth mass extinction, we are actually seeing the capacity of the Panama Canal being doubled. A Chinese investor even wants to dig another mega shipping channel across Nicaragua." I remind Joan of our previous conversation about Sicco Mansholt. How does someone in an ideological straightjacket rise through the ranks of the elite to reach the very top of political power and at the same time understand that an ecological economy is our only real option? Joan clarifies that Mansholt did not come to power as an ecologist but became one after he landed the job, especially after reading the well-known *Limits to Growth* report to the Club of Rome.[190] When published in 1972, it was the first major hard-science based study that showed with graphs and numbers that we were moving towards a total collapse of society if the economy continued to grow. Unfortunately, that is exactly what happened. The paper's predictions match the deterioration we see unfolding today. The collapse they predicted in 1972 by the middle of the twenty-first century is in fact right on their schedule."There is much more material today to have a contemporary Sicco that suddenly understands the real problem we have, also publications from the European institutions themselves", Joan argues. The European Environment Agency (EEA) wrote, based on a huge amount of data, the two excellent reports *Late lessons from early warnings*.[191] They notice that new technologies and products might bring surprisingly bad effects (asbestos, tobacco smoking) and that various crises are linked.

> Unfortunately, the EEA has zero political power. There are also people in the European administration that get the point, people who know that the current strategy does not work and will lead to mayhem. They are waiting in the corridors for when the political momentum for systemic change comes. We, you, me and many others must make sure that such political momentum will come. You do not have to be an Einstein to realize that you cannot solve a problem with the thinking that caused the problem. Do not believe the growth apostles. So-called green growth as a solution to our problems is what we have been trying for the past 40 years. It's what got us in this mess, not out of it.

When I steer the conversation towards the Paris climate agreement, Joan points to the lack of justice for the Global South. I ask him what he thinks about what the much-read environmental journalist George Monbiot said about the Paris agreement: "By comparison to what it could have been, it's a miracle. By comparison to what it should have been, it's a disaster."[192] Joan agrees with the latter.

> The ecological debt which consists of the damages inflicted upon the Global South is beyond question. In Paris, the world missed a unique opportunity to agree on a global tax on fossil fuels. Already in 1992, Italian environmental commissioner Carlo Ripa di Meana proposed to levy $10 on every barrel of oil. When he did not get it, he resigned. This simple but powerful measure needs to be put back on the table, especially now that the oil price is so low.

James Hansen, the NASA scientist who in the 1980s first raised the alarm about climate change, called the Paris agreement a fraud. He too says it is incomprehensible that no carbon tax emerged.[193] When the oil price rebounds, it will be much more difficult to sell a tax to the general public. The case for a carbon tax should be a no-brainer. The Belgian government estimates that a carbon tax of €40 per tonne would stimulate 80,000 new jobs, 2 percent economic growth, an increase in the available income for families and cleaner air.[194]

Another way to make fossil fuels more expensive is to disrupt the global supply chain. Joan's career was built on providing academic backing and political advice to environmental activists – mostly to those from the global South.

> For signs of hope, do not look to the Paris agreement. Look at Ende Gelände in Germany, where thousands of people stormed a coal mine and stopped production. Also look at Blockadia or the Idle No More movements in the US and Canada and the success of the activists stopping the Keystone pipeline. We should block fossil fuel extraction and transport and then explain well why we do it. Two Nobel Prize winners, Joseph Stiglitz and Paul Krugman, are on our side. Both are also against TTIP. They mainly use social

arguments. Krugman is against TTIP because it will destroy
small businesses. Even more interesting than that are the
arguments of ecological economists against TTIP, such as
those of Professor Alf Hornborg. He can argue TTIP away
with figures that reveal global ecological unequal trade. He
recounts our history in a very different way.

My school career went from economics to geography to conflict
studies. For me, Joan brings all these worlds together into a
single coherent theory on the state of play in the relationship
between humanity and the ecosystem we call planet earth – and
how we got here. His economics stack up a whole lot better than
the theories I was taught 20 years ago. It is all so obvious and
clear now. Neoliberalism in particular and the growth religion
more generally ignore geographic and bio-economic reality and
in doing so have created an increasing number of acute local but
also deepening global conflicts that affect every human being to
a greater or lesser extent.

    We should not have to argue about facts. But at a time when
politicians increasingly deny them, it has become necessary.
Post-truth politics has become a depressing phrase of our times,
one that is hard to avoid. Take Yanis Varoufakis. In his 6-month
term as Greece's Finance Minister, he tried to implement a policy
to rescue his country's economy in line with obvious facts. Facts
acknowledged by just about all major economists of our time,
including Paul Krugman, Joseph Stiglitz, Amartya Sen, Jeffrey
Sachs and Thomas Piketty. The issue they all agree on is simple:
Greece will never really recover without debt relief. But politics
denied this economic reality.Things get even worse when even
physical reality is denied. We have reached the point at which
a contemporary politician must be able to stand above the
laws of thermodynamics. More free trade thus gets precedence
over a climatological Armageddon that will reduce humanity
to a few tribes of cavemen plus a couple of hold-outs for the

superrich, if allowed to continue spiraling out of control. The substandard nature of most private media is creating a poorly informed majority who select representatives able to deny even the existence of gravity. After all, people keep joining the Flat Earth Society and they truly believe that gravity does not exist, that the earth is flat and that every proof that gravity does exist is just a ploy from NASA.[195] One poll showed that in France, 10 percent are now flat earth-ers.[196]

Facebook reinforces this mass ignorance. We consume and share the stories we want to believe without realizing that we are creating a walled garden, a lopsided universe where false news can spread faster than a rumor that Princess Diana is actually alive. And so it is that any Tom, Dick or Harry can get elected, based on fake news spread by people paid to advance a hidden agenda.

Private media following the agenda of their adverting sales team, social media echo chambers and merchants of doubt that invade both as well as the academic community; all of this is leaving us with our collective head in the clouds – far removed from reality on earth...It has led to the situation where the most trivial fart becomes a mighty thunderstorm in record time, with everybody chipping in their opinion based on rumor and fake news. During my student days I had a summer job at the Royal Meteorological Institute where I learned an important lesson: the higher in the stratosphere cumulus clouds get, the bigger and heavier the hailstones that fall to earth.

* * *

We are going exponentially higher in a countless number of ways but some of them were brilliantly summarized in a now famous peer reviewed paper titled The Trajectory of the Anthropocene: The Great Acceleration.[197] Will Steffen and his team put together socioeconomic and ecological trends from the last 250 years. All

12 socioeconomic and all 12 ecological parameters show, to a greater or lesser extent, some form of exponential rise. Sometimes older data is missing and sometimes the increase is more gradual, but it becomes obvious at a glance that from around 1950 onward, growth of almost everything became exponential. Where the impact on ecosystems used to be primarily local, it is now global. In sum, we are all together in one big boat today, one that is getting increasingly top heavy, unstable and yet going faster and faster. The question should not be if we should be throwing stuff overboard to stop our boat from going under, but what. Yet most policymakers and economists are still busy loading the boat, despite our peril.

A year after our meeting, Joan and I meet again, this time in the Hotel des Colonies in Brussels. The hotel took its name just after Belgium received the Congo from Leopold II, as it used to be his private playground. Under his brutal dictatorship, the population declined from 20 to 8 million. Neither Stalin nor Hitler came close to what Leopold inflicted. But Joan and I skipped the historical debt of Belgium to Congo in favor of a talk on private and sovereign debt.

At the beginning of October 2016, the press revealed that Deutsche Bank was on the verge of bankruptcy. This mother of all banks is much bigger than Lehman Brothers ever was. It is the most systemic bank in Europe, meaning its fall would have the highest ripple effect throughout the whole European economy. What would happen if the shit really hit the fan? I wanted to know from Joan what the ideal plan would be to turn such a mega crisis into an opportunity for radical change.

If a bank like Deutsche falls, that's the right time to go back to state-owned banks. They might even keep 80 percent of the money they are given by customers on their accounts, instead of 10 percent now. Debt repayment is always used as reason why we need GDP growth. But now that repaying all our debts is never going to happen, we have to be open about it. It is likely that you need to write off about half of all debts that exist today. You should read the book Debt: the First 5,000 Years by David Graeber. Defaults on debt are a regular and recurring event in history. Look, there are three ways to deal with the debt issue. Neoliberals would have us save up to pay off the debt. The green Keynesians would say that, at such a moment, you should invest more so that you can pay back by adding growth and then repay it all. Ecological economists say they are both living in a fictional world; that the debt is not repayable and that you need to have a major debt restructuring. The sooner the better.

If you dare to say this out loud, as Varoufakis did, the financial inquisition of rating agencies and the banks who pay them will be on you like a tonne of bricks. For the time being, most politicians rail against debt restructuring as totally absurd. Roland Barthes wrote in 1957 that: "Modern civilizations also have their mythologies."

* * *

A handful of rich businessmen are dreaming up some potty solutions. Billionaire Elon Musk wants to build a colony on Mars. Billionaires Richard Branson and Larry Page want to mine asteroids. Jezz Bezos is a space rocket man. Together they propagate a dangerous myth; that if we mess this earth up we can start again somewhere else – out there. In this blue sky or rather worm hole thinking, the earth is not a miracle where life is possible but something we can reconstruct elsewhere. The arrogance of those who think humans can live without water is mind-boggling, yet reels in hordes of supporters. When humans sense danger, they react in one of three ways: freeze, fight or flight. The people dreaming about going to live on Mars are probably of the flight-kind. But swapping planets does not differ much from jumping into a deep and dark abyss while just hoping for a soft landing. It appeals to the *there's-a-solution-for-every-problem* kind of people but it's just a distraction, a totally delusional and in fact dangerous idea – as it lures quite a few people away from even trying to solve the problems in this house that we call earth.

In a less outlandish manner, many more hold on to the vague belief that technological advances and efficiency gains will solve all our problems. Never mind that even the renowned Massachusetts Institute of Technology, a hotbed of technological advances, recently warned that technology alone will not save us. Demand for raw materials will continue to grow, despite efficiency gains, it confirmed.[198] And no mined asteroid is going to make up for the finiteness of the earth's resources.

The reality of our existence is rapidly catching up with us. Yet for too many, it feels easier to break the laws of nature than quit our unhealthy addiction to consumption, to stuff. With each passing year the writing on the wall is going to get more and more stark. Sometimes you can take that in a literal sense. The satirical graffiti artist Banksy hit the nail on the head when he defaced a blank billboard, scrawling an unsettling message: "Sorry! The lifestyle you ordered is currently out of stock."

The growth apostles, like so many religious leaders, prey on basic human needs for security, purpose and hope. They plug into such core needs of homo sapiens in order to sell a story that gives meaning to it all, despite defying all reason. To what extent do the apostles of GDP growth differ from the apostles of the Islamic State? In both cases, there is the promise of a paradise for those that follow the script, an unprecedented fantasy world that gives hope to those that feel frustrated, side-lined. The working poor that voted for Trump and those from the Paris and Brussels suburbs that went to Syria to fight probably share feelings of injustice, frustration and revenge. The mantras of their leaders are similar: if we just apply a little more orthodoxy to get to growth, or Allah, then your boat, or soul, will rise to paradise too! Charismatic leaders and their marketing gurus create a sense of purpose and a mythology and they both know how to use social media to reach their base.

Sometimes, the brain damage of the Growth cult really shocks me. *Pretpark Nederland* is a documentary on the outlandishness of consumer culture in the Netherlands. Viewers see a shopping center being passed off as an "authentic village" – to enhance the shopping experience. Nobody sleeps in the village at night, because all the houses are shops. In the docu we learn that one couple sleeps here: the shopping center manager and his wife. She puts on a happy face when asked what it is like to live in an artificial world. Without any hint of irony or humor, she happily proclaims: "Buying, buying, buying. That's what we were born for, no?" Buying is the new praying.

I think that deep down a lot of us "know" there's something horribly wrong with the average levels of consumption in the West but we tend to push the too uncomfortable thoughts away because if we let too many of them in, our heads would explode. The art of letting knowledge seep through while remaining sane, hopeful and if possible with bursts of happiness is one of the bigger challenges in modern day life. By reading until here, I can only hope you've mastered this art already.

Professor Tim Jackson summarizes the growth philosophy in his widely acclaimed book Prosperity without growth: "People are being persuaded to spend money we don't have, on things we don't need, to create impressions that won't last, on people we don't care about." His vision of prosperity without growth resembles that of Nobel Memorial Prize in Economic Science winner Amartya Sen and his idea of "development" as the capability to flourish – instead of the sum of all products and services delivered in a year. Jackson says that the capabilities to flourish are limited by the number of people on the planet and its finite raw materials. In short: the key question is how to get to prosperity for all within planetary boundaries. For that to happen, GDP will need to decline in some regions of the world, especially in Europe and the US. But that doesn't mean we need a calamity.

Tim Jackson and Professor Joan Martinez Alier are part of a fast-growing economic discipline that gives me reassurance, purpose and hope: the academic degrowth movement. In contrast to the cult of growth, this economic discipline is aligned with the laws of nature – such as the law of thermodynamics. According to Giorgos Kallis, sustainable degrowth – which stands in very sharp contrast to the disastrous degrowth his native country Greece has experienced in recent years – will see a fair reduction of production and consumption so as to improve human well-being and ecological conditions.[199] Riccardo Mastini defines degrowth as both spatial and generational justice. For me, degrowth is not about recession or austerity but mostly about more ecological quality, more resilience, more security. It's about reconnecting to nature and to others. And about doing what we still can to maybe prevent or at least postpone a full collapse.

Either way, the degrowth school is seeing a rapidly rising number of peer reviewed publications. The number of participants in degrowth conferences is also growing exponentially. My first degrowth conference was in Barcelona in 2010. Some 200 people participated. When I joined the Leipzig conference in 2014 there were around 3000 participants. It portends an economy of the future that looks completely different at local and at global level. We will see shrinkage in mining, production, transportation, consumption and waste sectors, replaced by growth of partnerships, ecosystem services, care for people and planet, resilience and happiness, localized agriculture and currencies.

Degrowth will see a drop in GDP, especially in countries where GDP is way too high to be anywhere in line with planetary boundaries. Instead of only innovating more and more technologies, there will be exnovation of technologies we can no longer afford – like fracking or the extraction of oil from tar sands. To get to the world we want, some of the things we do in the current world will have to go. This will sound disturbing to some, until we all understand that in this Great Transition, as some like to call it, we are switching from dirty to clean air, from loneliness and burn-outs to a thriving community life, from dependency to resilience. Professors Jackson and Alier are heretics to the growth church. Most so-called green politicians acknowledge in private that the pair are right, but rarely find the conviction needed to attack the growth orthodoxy openly. In 2016, Philippe Lamberts, a green member of the European Parliament, has pledged repeatedly that he will never call for degrowth. His platform? A degrowth conference in Budapest. It would be too repulsive. It is one thing for a relatively obscure figure to go against the grain. It is quite another for a prime minister, and we are a long way from seeing that happen.

But it does seem that the times they are a-changing. In early 2018 I co-organized a debate between degrowth economists, Philippe Lamberts and heads of units from several departments of the European Commission, including DG GROW. Not only did everyone seem to agree that we need to move beyond GDP growth, a poll among the audience – for the most part staff of the European Commission – showed nobody wanted the "jobs and growth" slogan of their commission. Instead, "People and planet" scored best.

In the autumn of 2018 I co-organized a major postgrowth conference in the European Parliament, with members of European Parliament from 5 political groups and European Commission representatives. The day before that conference I helped to get an open letter from 238 academics with postgrowth demands into mainstream media – front page – causing a national debate on the growth issue. Belgium's Minister of Finance felt the need to react and Balgium's largest trade union came out in support of the academics. The petition got 6000 signatures in no time, even though it is an academic letter calling for postgowth and not a call to save the last panda or whale.

Much still needs to happen, but a political ripening of minds is possible. A certain Mahatma Gandhi once said "First they ignore you, then they laugh at you, then they fight you, then you win". In 2018 this degrowth debate has gone to the stage where the vested interests are fighting back.

* * *

This possible momentum is hard to spot from outside a small bubble. Today, professors such as Joan Martinez Alier, Paul Verhaegen, Giorgo Kallis, Tim Jackson, Jared Diamond and Richard Heinberg are still like the floor staff told to man the shopping center doors on Black Friday. Their chances of calming the baying mob are slim. It does not help that their message rarely takes flight on Twitter, and is considered too negative for TV. And if it is not on Twitter or on TV, it might as well not exist. At least in the popular imagination. On the upshot: the number of degrowthers sure keeps growing.

When I am enjoying a more optimistic day, I share the opinion of Tine Hens. In her book Het Klein Verzet (The little resistance) she writes how for 30 years, Milton Friedman and Friedrich Hayek formed a stubborn sect within uniform economic thinking. They spoke a lot, but few listened. Then the oil crisis struck, the economic tide turned and powerful people started to embrace their neoliberal principles. The neoliberals went from crackpots to kings. Friedman later said of his wilderness years: "Only a crisis – actual or perceived – produces real change. When that crisis occurs, the actions that are taken depend on the ideas that are lying around. That, I believe, is our basic function: to develop alternatives to existing policies, to keep them alive and available until the politically impossible becomes the inevitable." Why would this be different for degrowth economics? In ecological terms, we are in a deep crisis. It might require just one more freak storm to bring degrowth alternatives in from the cold. The crisis is here and we are ready with the alternatives. What remains is something that makes the majority see the crisis that we're in for what it really is and then recognize the opportunity for positive change that we also have.

The fact is, the degrowth economists have been preaching for over 40 years. In 1973, German economist Ernst Friederich Schumacher published the ground-breaking book Small is beautiful. But small-scale islands of alternatives are not enough to turn the tide. Understanding has grown that a flourishing sub-economy is a step in the right direction, but in itself it remains insufficient to halt the collapse of our ecosystems that the much bigger players – like Big Oil – will keep bringing closer. As long as the sub-economy remains a niche of the GDP economy, it only marginally reduces the speed at which we are going to the wall. It only prepares a small part of the population for the shock to come. Hence, the attention has shifted from developing a niche and ignoring the rest to really attacking the mainstream with the aim of dismantling whole sectors of the economy that we just can no longer afford to have. As degrowth professor Tom Bauler told the people from the European Commission in our debate: "You have urgent work to do on exnovation, not just innovation." Degrowth economists ask themselves how an economy based on small-scale exploitation, fair trade, sensible consumption and a circular economy in both material and energy can work out well for all of humanity.

Sometimes, their thinking is picked up in the mainstream press. George Monbiot wrote the article *Growth: the destructive god that can never be appeased.*[200] He claims that the current pursuit of GDP growth is like putting the furniture in the fire to keep the house warm just a little longer.

The great demise of growth is coming pretty soon, either through an orderly victory of reason based on science or through the self-destructive inconsistencies of the current growth paradigm. The demystification of neoliberalism is a very sensible first step – but going from deregulated capitalism to old-school capitalism is not the solution to our problems. There are some hopeful signs that neoliberalism is being attacked, but often what comes in its wake is not what we want. Rather recently, Europeans were subjected to the neoliberal policies of Tony Blair in the UK, Gerhard Schröder in Germany, Louis Asscher in the Netherlands and François Hollande in France. By forgetting why they were founded, they and the big old socialist parties they worked for signed their own death wish by going down the neoliberal path. But today, the opposition to neoliberalism seems a lot stronger. In the UK two-party system, Jeremy Corbyn went back to the roots of Labour but in most other countries the parties left of the now neoliberal socialists have grown strongly – for example with Jean-Luc Mélenchon in France. A left-wing Portuguese government steered the country to recovery in a manner starkly in contrast to the austerity seen in Spain. Anti-neoliberal progressive parties are recapturing the leftist void that the mainstream socialist parties created for them to step into: Syriza in Greece, GroenLinks in the Netherlands, the workers party in Belgium (PVDA-PTB), the list is endless and the pattern is clear. The big old socialist parties in Europe are learning the hard way that there's no voter base for neoliberal socialism.

The problem we now face is that when the global neoliberal order is the house, more people flee through the right than through the left door. Marine Le Pen, Nigel Farage, Geert Wilders and the likes have been better at turning frustrations with "the system" into votes. In their world view, the biggest victims of globalization – those who need to leave everything behind in search for a life worth that name – are quickly rebranded as the common enemy against which the troops can be rallied. But the problems faced by many of the Western people who vote for them does not originate in some remote conflict-struck country, it originates right at home. In 2016, George Monbiot wrote what became the third most widely distributed article published by The Guardian that year, with over 700,000 shares: *Neoliberalism, the ideology at the root of all our problems*. Monbiot smoothly linked neoliberalism to the financial crisis, the ecological crisis, the unemployment crisis, the rise of loneliness and the rise of Trump. That was in April 2016, before Trump's election. After it, the media was full of articles explaining how neoliberalism had undercut a generation or more and swelled the ranks of Trump's base. It is worth observing that it is in the US rust belt that the elections dealt the greatest pro-Trump switch – which came as a major surprise to almost all the pundits and political commentators. These are the states that suffered most from neoliberalism, with loss of income and jobs leading to ghost cities.

The time to make these connections plain and obvious across the media is long overdue. Understanding the connection between neoliberal failures and the rise of fascism is a necessary step towards abandoning some twentieth century myths that threaten our very existence as a species. The way out of this mess is not to look for failed economic system ideas from the past, but to look forward to the twenty-first century. Keynes, Mao and Friedman all had one god in common: Growth. We simply cannot afford to sacrifice everything on its altar any longer.

* * *

But how to bring down a religion so pervasive it has become invisible? Some numbers are helpful in this struggle. Today, there is a scientific method to calculate the day we cross the threshold from sustainable to unsustainable consumption of natural resources for that year. After passing Earth Overshoot Day, we eat into our natural capital, making it harder to achieve sustainability the year after. We use some of the stock of natural materials formed on earth over the last 4 billion years. In 2018, everything we took from earth after 1 August was too much. Back in 1970 everything went well until 29 December. The sustainable year is getting shorter and shorter every year, and is simply inconsistent with the policy of pursuing perpetual GDP growth. How to push back overshoot day is a difficult question. But the real question is what economy we need to make sure it stays in the green zone until 31 December. That economy will no doubt have a whole lot less material flowing through it.

We must do that because the consequences of not doing it have also been quantified in great detail. The University of Melbourne judged in 2014 that the Club of Rome's 1972 report was mostly correct.[201] They confirm we are on track for a collapse of global civilization within this century. Other research shows that a policy aimed at economic growth is the driving force behind the long-term destruction of the conditions that make life on earth enjoyable and that without radical adjustments, the collapse is predicted by the middle of this century.[202]

Every year the growth apostles get together in the Swiss ski resort of Davos. A recent report presented in Davos zoomed in on how the growing inequality in the world is the greatest danger to continued GDP growth. Read that sentence a second time. Finishing with "Amen". When will there be a report in Davos about how the focus on GDP growth and the neoliberal policies that are associated with it are the greatest cause of rising inequality?

Chapter 1

The heresy of a pope

Humans need a story that inspires. Given the state we're in, it's not evident to bring one while doing justice to the facts. Whether any hope is justified when put next to the massive problems mentioned earlier, I honestly can't tell for sure. My judgment on that also depends on the amount of the sun, the quality of my coffee and the number of hours I slept.

As Alex Evans explains well in The Myth Gap: Stories have far greater power to move people than facts. Think of Boris Johnson touring the UK with his 350 million £ bus or Trump saying that the wall just got higher. To compete with their fake but hard to resist believing-in stories, we need charismatic leaders that inspire people to confront the issues I've described. Enter: Pope Francis.

The current leader of arguably the greatest myth reproducing machine humanity has ever seen is a myth-buster. I declared that way before Pope Francis declared that there's no such thing as hell. When starting a chapter on hope with the pope I must say I'm neither catholic nor the kind of guy that likes religious leaders in general. But there's no point denying that this Pope is a reason for hope.

His choice of papal name, Francis, was a warning sign. According to the Franciscans, there is an inseparable link between inner peace and concern for nature and the poor. But even then few had expected such a radical ecological text as Laudato Si, which he published in 2015. It shook the foundations of both his own church and the church of Growth.[203]

What follows next comes straight from what I'll refer to as the Earth Bible. You better buckle up. Pope Francis rails against our waste culture, the idea that technology will solve everything, overconsumption, privatization of the water supply and public spaces, fracking, the use of cyanide and mercury in gold mining, the excessive use of chemicals, industrial tree plantations, dynamite fishing and a few other practices that are pillaging our ecosystems. In his own words, Pope Francis denudes "the idea of infinite or unlimited growth, which proves so attractive to economists, financiers and experts in technology. It is based on the lie that there is an infinite supply of the Earth's goods." Hallelujah!

He recounts how everyday experiences and science have taught him that environmental destruction hits the poorest the hardest. The term environmental justice was coined by black communities living in the shadow of a landfill that happened to hurt the black community a lot more. Today, the term remains relevant but is heard less often. That is because, according to the Pope:

> many professionals, opinion makers, communications media and centers of power, being located in affluent urban areas, are far removed from the poor, with little direct contact with their problems. This lack of physical contact and encounter... can lead to a numbing of conscience and to tendentious analyses which neglect parts of reality...We have to realize that a true ecological approach *always* becomes a social approach; it must integrate questions of justice in debates on the environment, so as to hear *both the cry of the Earth and the cry of the poor*.

But Francis is far from done with that. He then lashes out at the continued use of harmful production methods in poor communities that were long ago deemed unlawful in the West. He points out our glaring ecological debt, racked up from years of over-exploitation of natural resources. The Pope's rescue plan? A significant reduction of consumption in countries that did most to put us in the red. He has been critical of global environmental agreements that are weak and focused on technological fixes and financial hocus-pocus.

He then takes aim at the classical position of the church in society. In the last centuries, the church was mostly on the side of the capitalists. Both that church and the capitalists believe that humans are above the rest of nature and they can therefore control nature, be the masters of it. By presenting the earth as a sister and mother, Francis shreds this centuries' old view held by the church. At the end of his Earth Bible, he proposes a kind of ecological conversion to ecologically aware citizenship, not just of the 1.6 billion Catholics on the planet, but to all its inhabitants. Bottom line: we can no longer remain passive in the face of the destruction of the ecosystem.

based on the assumption that humans are all selfish and that homo sapiens is homo economicus. Neoliberals think that the economy is above nature's laws. Neoliberalism and the laws of thermodynamics just don't go through the same door.

But the key question is not whether degrowth academics are masterminding yet another ideology or not. No matter how you want to name the fast-growing community of degrowth academics, the fact is that like neoliberals, they too see things big. They do not fall into the trap of believing in labels and complex financial products as a means to hide the real causes of our problems. They stopped denying what most economists still deny: green growth capitalism won't save us from a collapse.

The geophysical scientist Brad Werner studied this in depth before coming with a presentation to his colleagues with a hard to misinterpret title: "Is Earth F**ked? Dynamic Futility of Environmental Management and Possibilities for Sustainability via Direct Action Activism." In his advanced computer model there was only one variable that offered hope that a collapse could still be avoided: the anti-capitalist resistance movement through direct environmental action, blockades and sabotages.

As Naomi Klein rightly concluded from that presentation: only mass social movements can save us now. The anti-capitalist academic degrowth movement provides the intellectual foundation for the global movement for environmental justice to do what's needed – even if laws need to be broken. Naomi knows how big the challenge is: "everything is easier than imagining how to change capitalism, but it is necessary to prevent a catastrophe".[206] She illustrates it with how we as a human race fail catastrophically when it comes to facing the climate crisis. Going from all the meetings and agreements we've come up with, you could be excused for believing that we are in the process of controlling the problem. But in practice, this is clearly a scam. Since the first climate negotiations began in 1990, emissions have risen by 61 percent. Klein also sees the creation of a global free trade regime, in parallel with climate negotiations, but capable of meting out fines worth billions, as a crucial reason for our failure. She makes a convincing plea for growth in the quality of care for each other, while shrinking our use of raw materials in absolute terms.

So, where do we start? Well, maybe we should first change the compass of the ship we're all on. To measure progress – and most people do want some measure of progress – we need new standards. The alternative narrative will also be about growth, but of a very different nature than growth in mining, transport, consumption and GDP. The Genuine Progress Indicator (GPI) is one example. It corrects the GDP. It deducts issues such as use of natural capital, pollution, crime and increasing inequality from the benefits.[207] Home caring and volunteering are included. In total, there are 17 corrections to GDP to better measure real progress. The GPI measures real progress in the journey of humanity on earth.

This is what happens when you measure progress using GDP: If you put all countries on a scale from poor to rich, you will see a perfect correlation between GDP and GPI – until around $7000 GDP per capita, per year. But in countries with a higher GDP, the correlation totally disappears. To the contrary: while the world's GDP has tripled since 1950, global GPI has fallen since 1978.[208] In other words: while GDP keeps growing we are actually reversing the progress we made until 1978.

Of course, as William Bruce Cameron wrote: "Not everything that counts can be counted, and not everything that can be counted counts." But to circumvent the focus on the GDP and deal with the human weakness to translate everything into figures and tables, you probably need an alternative indicator, one that can grow without wrecking the conditions for life on earth. It also seems obvious that there's no one-size-fits-all solution. It may be good for a village in India to use the water table as their key progress indicator, good for Bhutan to use Gross National

Happiness (more on this later) and in the West to use GPI. Each indicator has its weaknesses and strengths and its relevance will sometimes depend on regional geography or history. However, there is no real reason to hold on to an indicator like GDP that does not succeed in measuring the difference between more ice cream or more crime. We all want more of the former, but not more of the latter, right?

More crime equals more GDP. In the US, where the prison system became privatized, GDP rises when they put more people in jail. Accompanying a trend towards privatization has been a rise in the number of prisoners, increasing by 600 percent in 4 decades. One in four prisoners worldwide are US prisoners.[209] The two trends show a frightening positive correlation. In a country where GDP growth is worshiped, the correlation makes sense. After all, once you have a hotel, you want it to be full, no? Once you have private prisons, laws become stricter and more people are sent to prison. Sometimes a more forceful approach is taken and judges are bribed in a kind of kids-for-cash program. Judge Mark Ciavarella was paid by the private jailers to incarcerate children. He obliged in the cases of 4000 children brought before him.[210] In the GDP economy, jailing kids counts as progress. It is efficient. If there are problems with other indicators, they need to be seen alongside the failings of GDP as an indicator, of which bribing judges to put kids in jail is just one example of what counts as a plus in the GDP.

The word degrowth scares many, but I'm scared by what falls under our key progress indicator today, which GDP still is. Those wishing to avoid controversy shy away from naming it openly, preferring other terminology such as post growth or the related but softer steady state economy. But it was Adam Smith who said that, in the end, growth would no longer be needed. Other macro economists such as Silke Helfrich prefer to ignore the GDP indicator altogether and rather study the world outside of the market and state.[211] Helfrich was a keynote speaker at a degrowth conference in 2014 that brought an eclectic set of resistance groups from the global north and south together in Leipzig. In the global north, you have academics and transition town groups. In the global south, you have a lot of people who do not call themselves environmentalists, because it is about injustice and survival for them. In the degrowth movement, those previously disconnected worlds come together under one umbrella.

Whether it is the Pope, professors of ecological economics, writers like Naomi Klein, environmentalists from the global south, a movement of movements in Belgium or the author of The Myth Gap, they have one thing in common. They all say we need a new inspiring narrative, a story in which the fruits of the earth are shared a lot more fairly. A lot of them identified a crucial tipping point: getting the trade unions on board of the environmental struggle.

But how can unions ever turn against their own industries? This feat of persuasion reached some level of success in England, where a big union in the oil sector signed a memorandum calling for the oil sector to disappear in the next 30 years. How? By first talking about jobs within a socially just degrowth economy. Jonathan Neale of the UK 1 Million Climate Jobs Campaign said: "Every job that disappears in the oil sector needs to and can be replaced by a job in renewable energy – if we make energy jobs public jobs." Work-life balance, continuity and appreciation for labor are the reasons why Neale wants public jobs. He also argues that energy supply is too important to leave to the private sector. "These renewable energy jobs are healthier jobs that give workers a better feeling, a feeling that they are working on something positive. Once the militants were convinced, the union leadership had to follow."

In the same Leipzig conference where I spoke with Neale, Professor Willi Haas told me something else about work that opened my eyes. "Politicians always say we need GDP growth because we need jobs. But why don't they create the jobs directly?" Does that relationship even still exist? One study showed that there is no longer a clear correlation between GDP growth and job growth.[212] The report states: "For these authors of very different persuasions, there is no need for short-term growth to create employment. The need is more for policies that boost...employment."

The struggle of the people who want to save workers and the struggle of the people who want to save our ecosystem are one. Neither workers nor ecosystems need GDP growth. People need justice and security – and depend on the quality and abundance of nature. We need to grow in care for one another and for nature, not in GDP. Some will do everything they can to divide and rule, to place the worker against the environmentalist, but in reality their struggles are one.

The work of civil society is too often divided. While unions protect those with formal employment, non-profit communities are most often split between environmental, social, gender and so many more issue silos. Thousands of organizations exist with as their core business the fight against poverty and inequality but when working in their silo only, they are fighting an uphill battle. By 2014, the poorest half of the world's population had the same level of wealth as the richest 62 people. In 2016, that went down to the eight richest people.[213] Thomas Piketty showed that this trend will continue in the twenty-first century unless we take action to tame the so-called free market.[214]

Degrowthers also argue that, in addition to the reduction of extreme poverty, we have to look at the reduction of extreme wealth. Some have developed concrete proposals, such as legally enforced pay ceilings, so that bosses are paid no more than five or ten times that of their lowest paid employees. Progressive banks like Triodos exist where such maxima are applied already. The barriers to achieving a fairer world are political, not technical. It is not about natural laws, but about the choices we make.

The toolbox is there. It's all about creating the momentum to be able to open and use it, about saying *Podemos* with a majority of people.

Chapter 3

From neoliberalism to environmentalism

In This is tomorrow, Thomas Decreus and Christophe Callewaert write that neoliberalism went into intellectual bankruptcy in 2008 and that we now live in an ideological desert.[215] But even in the desert, one shock rainfall can suddenly give rise to a multitude of beautiful flowers. In Latin America, people talk about *buen vivir*. In Bhutan, the key measure is gross national happiness. In European academic circles, the degrowth narrative is not the only but one of the bigger stories behind well-targeted direct actions against the most polluting mines. We do not need a one-size-fits-all approach for everyone. History, geography and cultural values all play a role in why one model works in one area and fails to strike a chord in another.

Sometimes intellectual flowers bloom globally. Certain technologies unite people across all borders. Wikipedia is a well-known example of a truly global peer-to-peer online network. In this case, the niche eclipsed the mainstream Encyclopedia Britannica. The internet also brings together activists who otherwise carry an isolated "David against Goliath" fight against a big company. Today, action groups exist around the world that act locally against the malicious practices of companies such as Chevron and that gain strength from exchanging information with other action groups online. At shareholder meetings, they act en bloc and in some cases they force mining giants, like Vedanta, to their knees.

Proponents of business-as-usual like to paint degrowth proposals as Stone Age ideas. But the people ridiculing degrowth often have a self-serving motive when they laugh at *degrowthers*. Degrowthers don't even want to go back to the seventies, when our ecological footprint was still manageable.

Take the smartphone, which didn't exist yet. Even in a degrowth society, it is okay to use a smartphone. The difference is that it is going to be something like a Fairphone or something you lease from a company with an interest in ensuring the phone lasts. You are probably going to charge it with something like a WakaWaka Powerbank: a tiny portable solar energy panel. There will not be a billion-dollar marketing campaign around each new version and there will be no new version every year. There will be upgradable parts that people can change themselves. But the functions of the smartphone itself will be the same. It's not about going tough, giving up or being nostalgic. It's about getting rid of the things we don't need: billion-dollar marketing campaigns, mega-shops to sell tiny phones, super-bonuses for managers and planned failure of products – to name but a few things the world can sure do without.

Degrowthers are not against consumption, but they do put the security of a system that everyone needs above the individual freedom to over-consume. Overconsumption deprives others of the freedom to consume. Freedom takes on a different meaning in a degrowth society. Take car ownership. "My car, my freedom", so goes the cliché. But the "my" in "my car" implies a recurrent purchase, insurance, tax, maintenance and so on. In Belgium, the average car owner sinks a fifth of their income into their car. Put another way, they work one whole day each week to be "free". In Dutch, Friday means "free day" – but in reality it's rather the opposite: most people will be working on "free day" to be able to *own* a car – rather than simply using it…and be free on Friday.

Is car ownership the best way to get around? The number of people who think so, who own a car, is in decline. They still use cars, though. These are usually people that live in cities, where alternatives to the car have multiplied. In the rare cases it makes sense to drive a one-tonne vehicle to transport their 70-kilo self, they will simply use or share a car from Cambio, Bolides, Tapazz, Dégage, Uber or, well, there are now too many options to name them all, but in the main cities there is something for everyone. Not only do they offer much cheaper transport, they slash the number of cars produced, and the road and parking space needed for cars. On average, a single shared car replaces 14 individually owned cars. There is also a snowball effect, with a growing membership allowing car schemes to set up more parking space which in turn attracts more members. Today, mayors all over western Europe are starting to realize that reining in king car also means giving back the public space in the city to the citizens. Soon we'll look back at the age where cars totally dominated the public space in cities and think: whose idea was that anyway?

Freedom in a degrowth society means that nobody expects you to work 5 days a week. Part-time work as the default option will make you rich in something that most people complain about not having: time. Time can thus be allocated to living more sustainably and to avoiding one of the fastest growing diseases of our times: burn-out. In Belgium, burn-outs doubled in 5 years. Sick leave days rise exponentially. The pressure on full-time workers is just way too high.

And it's true: all that cycling, community farming, repairing and sharing of stuff takes some time. But who says that working 40 hours a week is best for a thriving society? In fact, a 21-hour normal working week is far closer to the mark, according to the New Economic Foundation, an influential British think tank. It would relieve the pressure on an overworked people, the authors say. Associated medical problems would be improved, as would unemployment, overconsumption, environmental harms. Our well-being would be lifted, as would equality and the amount of time available to care for others and generally enjoy life. The foundation makes a convincing case for a:

> substantial reduction in paid working hours, aiming towards
> 21 hours a week as the norm. The current norm of a nine-to-

five, five-day week in paid employment does not reflect the way most people use their time. Unpaid work is generally overlooked and undervalued. A much shorter working week offers very considerable benefits to the environment, to society, and to the economy.[216]

This is not some kind of New Age fantasy, but a realistic roadmap to the low carbon economy. Nobody in the UK regards the New Economics Foundation as an out-of-touch hippie hangout. There's macroeconomic progress put into practical policy advice for a degrowth society. It just makes a whole lot more sense than the out-dated twentieth century 8-8-8 principle: 8 hours at work, 8 hours free and 8 hours of sleep. In the twenty-first century it is about time we go for a 50-50 principle: 50 percent of working time in the formal and thus by default global economy and 50 percent in the informal and mostly local economy. In the latter a local currency is used to exchange goods, repair, practice city or community agriculture, care for others etc. Of course, a 50-50 principle would result in a smaller GDP but most likely also a higher GPI and a higher personal well-being for most of us.

One of the main hurdles to this revolution in work is that at present, employers are usually not keen on it. I was lucky that my bosses were open minded enough. For years, my 3 days a week in the global economy gave me the space that I need to also write articles and books, whenever I feel like doing that. But I know that most people just do not have such understanding bosses and therefore I believe it is up to governments to make a 50-50 principle the norm, of course with some level of flexibility. Those who really want to work more in the formal economy should still be able to do that. But changing the default makes the most economic, social and ecological sense.

The first hurdle to get there, however, is planted deeply in our own brains. The dominant psychological climate is all about being "ambitious" and ambition is translated as working hard for a single employer. Loyalty is confused with fidelity – as if you are cheating if you're also working for someone else. Professor Paul Verhaeghe says that 30 years of neoliberal policy has had a profound impact on our personalities and on the profile of a good employee. He says that to build a career, you now need the characteristics of a psychopath. But there is both a push and a pull for this psychological climate. Jobs have become less secure, so we get inclined to conform more to the demands of employers. At the same time marketing gurus and advertisers tempt us into buying ever more stuff. The only way to square the circle is through more work. We have to break through that vicious circle.

Another myth standing in the way of a shorter working week is the riddle that you can only think of the environment once you are rich enough to afford such luxuries. Think about buying more expensive FSC wood and a Tesla. Neither of that will save the world, to the contrary. A really sustainable lifestyle makeover entails things like recycling wood and ditching a car for a bicycle and car sharing. Going green usually saves you money, no matter what the ads tell you. Nowadays, you can insulate your house with shredded newspapers, buy decent second-hand clothes or pay a farmer a yearly fee so you can go to his farm and harvest local organic food whenever you want – and save money while doing so. Some cooperatives for renewable energy offer electricity at a lower price than power from the dinosaur firms still offering fossil fuel energy. The thousands of people behind Ecopower decided to give low-consuming customers a special discount, thereby sending a signal that people who use far below average in their home get a reward. You do not need to be rich to be green. Living sustainably is both rewarding and liberating. It delivers a good life of less stress, less burn-out and more quality time. It's the road to a more social kind of freedom.

Aside from the concept of freedom, the idea of "security" also needs a serious rethink. As written earlier: the level of security provided by capitalism is summed up in the phone calls from London city bankers to their wives to rush to the supermarket and stock up food, on the day Lehman Brothers went belly up. As the root causes of this global financial instability have still not been addressed, a global financial heart attack seems a matter of "if, not when". At some point, probably very suddenly, most people will be faced with a life-changing question they have probably never had to ask: How am I going to feed myself and my family now that the supermarkets are empty?

Europe is the most import dependent of all continents bar Antarctica. On the positive side, local supply chains do seem to be blossoming. Community supported farming is one example, but there is a whole lot more happening. Today, we can order online from nearby farms and eliminate the horribly long supply chains clogging our roads and threatening our climate. Local food chains can better withstand the kind of financial crisis that keeps bankers awake at night, all the more so if local currencies are involved. And you know where your food comes from, something less trivial than it may seem. The fact that more and more people are opting in to this form of grocery shopping is also evident from another Belgian initiative, Boer'n Brood. It started with 25 food fiestas in the city of Ghent, pop-up parties on urban wastelands where local farmers and townspeople assembled huge picnic buffets. The movement started small, with volunteers. But soon there were 20,000 people acquainted with products and producers from their home region. Now that is what I call security. Food security.

This kind of economy is also more transparent. To cater to the need for transparency, The Food Hub in Leuven is a place where ecological agricultural cooperatives sell their products in a pioneering transparent way. Each product is sold with a circle diagram showing consumers how much money goes to the farmer, how much to transport and how much to the shop. You can also see the distance that the product had to travel. Large posters with faces of farmers and a little bit of their story allow customers to get acquainted with the people that made all of the products sold in the store. The Food Hub wants to limit the number of food kilometers, make a connection between consumer and producer and go beyond the bio label – knowing that start-up ecological farmers can't get the label. One more food store that will not break down if the global financial system suddenly freezes.

But it is not just food where local supply chain options are on the rise. Almost every self-respecting city in Flanders now has an occasional flea market where all items are given away for free. It offers a space for those that are drowning in stuff to give away some of their things and in the process make others happy. There are now repair cafes where volunteers fix your broken appliances, also for free. There is the rise of Freecycle: Facebook groups where people just give things away. This economy does not register on the radar of the GDP economy, but it is very tangible and real. And it seems to be growing, at least where I live. Another example of this parallel economy is the one where goods and services are paid in local currencies. Our family joined LETS Leuven, a totally virtual "currency" used by hundreds and sometimes thousands to exchange things and services. This could be repairing clothes or cycles, showing how to make an online photo album, exchanging books or toys, giving a helping hand at parties and so on. When all money traffic stops from one day to the other, this local economy will be unaffected. In this safety net, the real currency is trust, not money. As this network matures, diversifies and grows more popular, we will be able to call on more and more back-up options like this.

There are so many more examples of individual steps towards more ecological freedom and security, but this book is not a do-it-yourself guide to the green living. Others are way more specialized in that, from Belgium's "low impact man" Steven Vromman and the five popular blogging ladies who wrote

Greetings from Transition-land, to Rob Hopkins of Transition Towns fame in the UK, or Pierre Rabhi from OASIS in France.

These glimpses into the parallel economy merely serve to illustrate that TINA is truly a myth and that TAPAS is a reality: There Are Plenty of AlternativeS. True environmentalism is not about impoverishment but about empowerment. It is through environmentalism that one becomes "rich", not the other way around.

Chapter 4

From free to fair and sufficient trade

All of the above lifestyle choices are not going to save the world, but they can give you, your family and your community a lot more freedom, security and protection. For me, participating in the solutions economy and fighting against the destructive myths of our time are yin and yang: they provide a sense of psychological balance. The engagement with groups that work on the here and now solutions on a small scale gives me the positive energy that I need to cope with my anger over the bigger problems and to drive negative energy into deconstructing the great myths of our times.

The market, state and the commons are the corners of a great triangle. The market is constantly trying to overtake the economies of the state and the commons, driven by neoliberal dogmas. In my opinion, both these trends need to reverse, without erasing the market economy completely. Groups of citizens need to claim the commons again, for example through renewable energy cooperatives. States must reoccupy some core tasks again, like banking. The operating space for a market to steer economic activity should be bound by stricter rules to protect what's good. Decreus and Callewaert write about the need for a bigger commons economy: "If we ever want to develop a political and social alternative for the current order, then we need to do everything we can to prove that those alternatives work better. And that is exactly what the commons economy does." I believe that a country can thrive sustainably if it has a state bank, a couple of cooperative banks and a few casino-capitalist banks – but that it's really hard when you only have the latter who decides where money can and should be invested in.

That second reversal is also needed. States need to take back at least some of the control they lost. States should prohibit planned obsolescence, the pre-planned failure of printers, phones, computers and so much more, just after expiration of the warranty. It should be a zero-tolerance issue, cracked down on as an environmental crime. The corporate executives who planned to rip off consumers and the environment should end up in jail.

Ideally, states will cooperate at intergovernmental bodies like the EU or at a strengthened United Nations. The pressure to do this is already there. Just one example is the global campaign to recognize ecocide as the fifth crime against peace, combined with the creation of an international environmental court under the wings of the current International Court of Justice in The Hague. That campaign was probably one positive factor in the surprise decision by the International Criminal Court (ICC) in September 2016, when it announced that it is also open for business when it comes to environmental criminals. While this is a step forward, the ICC is also a textbook example of too little, too late. It's not the ICC that polices the world. Truly ecological states, urged forward by their citizens, will need to be willing to hand over their environmental criminals to the ICC.

In this highly globalized world, there must be a stronger role for cooperation between states. The countries that make up this world have already proven that they can deal with pressing environmental issues. There was less than 10 years between the final determination of the problem associated with CFC gases and the signing of the Montreal Protocol, which quickly prohibited the production of these gases. Banning them has allowed what was a growing hole in our ozone layer to largely heal.

But if, for some reason, not all states agree to tackle, for example, climate change, there is still plan B. The US has in the past used the World Trade Organization to force Thailand to use turtle-friendly fishing nets. According to the prominent US economist and Nobel Memorial Prize winner Joseph Stiglitz, the very same logic is perfectly applicable to force the US to produce carbon dioxide-friendly products. Europe, Japan and other countries that signed the Paris Agreement could restrict, tax or ban the import of climate destructive goods from the US by applying World Trade Organization rules. In his book Making globalization work (2006), Stiglitz writes that the dangers to the world that global warming causes are certainly important enough to risk the displeasure of a villain state that is prepared to endanger the well-being of the world simply to maintain its vibrant life-threatening lifestyle. Stiglitz was pointing to the US as a villain state, long before Trump came to power and announced that the US would withdraw from the Paris Agreement. If we do not make Stiglitz's proposal work now, when will we?

Tighter product standards, prosecution of CEOs, such as Chevron's, and using World Trade Organization rules to ban or tax the imports of products from climate cowboy states; the net result of all this would be a reduction in the harmful part of world trade and an improvement in the quality of our ecosystem. What we need is a European government that can guarantee our competitiveness through positive discrimination on an ecological basis, and not by pushing down our wages. Limiting, taxing and prohibiting imports of those products causing grievous environmental impacts is a matter of political will. Everything else needed to get this done is already here.

One big problem we face is that trust in the global and European institutions is not exactly at historic highs. The reasons for this are no mystery. Our inter-governmental institutions claim too often that there are no choices beyond austerity. The IMF does not care who Greek people vote into power, as long as they do exactly what the IMF tells them to do: implement reforms which, in private, even they admit are not working. Institutions like the IMF, the ECB and the EC, which make up the trio of Greek creditors, simply want a technocrat in charge.

Stiglitz reacted to this saying that yes, certain issues must indeed be delegated to technocrats, such as selecting a good computer system to implement public pension payments. But setting the rules of the economic game should only be delegated to technocrats when there is only one single best set of rules, a set of rules that make everyone better off than any other set of rules and as Stiglitz said: "That's simply not the case." Which is why democracy needs to correct economic decision-making if the wrong choices are made.

Supranational institutions, ideally the UN, could ban food speculation, levy a redistributive tax on financial transactions, stop biofuel subsidies, create a global carbon tax and force ship owners to use cleaner fuels. One can develop an alternative indicator that measures growth in well-being. There is so much that politicians could do, if only they could break with the stupid dogmas that stop them.

At the Belgian Pavilion of the World Exhibition in Milan there were some beautiful examples of the circular economy in practice. A Brussels-based company demonstrated how coffee grain leftovers from restaurants can be used to grow mushrooms on, which they then sell on to the same restaurants. In Europe, Wallonia is the first region to embrace the circular economy as the new industrial policy. The Walloon resistance to CETA is more than a stubborn "no". It has a strategy to move forward without an increase in world trade. This futuristic Walloon vision of trade agreements is also reflected in the Declaration of Namur. Forty prominent academics from all over Europe signed it, including Thomas Piketty.

But there is also hope that the EU is taking the drive towards a circular economy seriously. After trashing the package of measures at the start of the new EC led by Juncker, someone has had the courage to stick their head in the bin and figure out a way to upcycle this scrapped package of laws. It could very well be that team Juncker has done this re-tabling of circular economy regulation out of economic rather than ecological motivation. After all, research shows that European companies could save €460 billion a year under a more circular economy. By 2025, this would create between 500,000 and 860,000 new jobs. Janez Potocňik was the former European Environment Commissioner who broadened the original package of measures. He said: "At present, we use 80 percent of all that we produce only once. We cannot afford that kind of waste anymore." The only sensible alternative to more world trade through TTIP, CETA and so many other agreements is a much higher degree of reuse and upcycling, one driven on clean energy. This is not only necessary for the global environment, it also brings back lost jobs.

Chapter 5

Pioneering companies

In the triangle of market, state and commons, the role of the last two is undermined and undervalued. This is not to argue that markets and private companies should not exist. The model where the state owns all companies has failed and while the commons economy can scale up I seriously doubt it will be able to fully replace all market activity. With that in mind I set out to find a promising market player, without putting blinkers on. After all, plenty of books have been written and documentaries made about promising cooperatives. The cooperative model seems to work fine for many small and medium-sized businesses, where people who know each other take all the big decisions together, including money matters. A company with 100 employees can usually run this way too. The criticism of ecologists and degrowthers is often that they have swung the pendulum too much and think on too small a scale for meeting the needs of soon 8 billion people.

I accept that this criticism is sometimes correct. Sometimes, a long and complex chain from raw material to product and back can be more sustainably organized by one large company. Such a company would need people at its helm that think long term and approach things holistically. But for all the positive practices of large multinationals, there is usually a flip side within the same companies. The same is true of the way countries are run. Since many companies are as big as countries, classifying either as good or bad does not do justice to the complexity and diversity of the practices within them. There is a company I know quite well from its promising practices. But I will not name it, because to do so would instantly form associations that narrow most readers' minds to the cliché about that company. Other people will focus on the dirty laundry of this company. In this less critical chapter, I will focus on the practices that illustrate to me that a large private company can also have a positive impact.

One of the many professional "waste pickers" of this multinational found a goldmine in the garbage bin of another company: a ski stick manufacturer. After cutting their sticks to get a diversity of ski stick lengths, a steady stream of short, medium and longer leftovers ended in the dump. The same ski stick company also generated a waste stream of plastics, after cutting circles from squares to make these round things you have at the end of ski sticks. Turns out that with these two materials, plastic and metal waste, you could make a perfect sleekly designed candlestick – which the multinational I've investigated then put into mass production.

This is just one textbook example of a win-win. One company no longer needs to add to the waste pile and the other does not need to expand the extraction frontier. My mystery company is actually a veritable Willy Wonka of the upcycling world, remaking things on an industrial scale. Think garden furniture made of used car tires, pillows stuffed with their shredded catalogs and home furniture made from car packaging – yes, it exists.

Yes, we have a lot of problems with monopolies and corporate giants that steer us in the wrong direction. Yes, we need to reverse that. But a very big company can also push the market in the right direction. Its head of sustainability explained how it did this with LED lamps, which last a good deal longer than any lamp before them.

When LED came on the market and we understood how good these lamps are for the environment, we jumped on them 100 percent. If you continue to offer energy-saving lamps that are twice as cheap but that last ten times less, then people will continue to buy energy-saving lamps. That's bad for the environment and in the long run for the consumer too. When we went for 100 percent LED in our shops, the world price of LED lamps decreased, due to our market share.

For cotton, this company wants to achieve the same effect: by going for 100 percent sustainable sources, it wants to push the price of sustainable cotton down. Labels are sometimes counterproductive, as I explained earlier. But what distinguishes this company from many others is that its boss is thinking about sufficiency. He says we are not just looking at peak oil but at peak stuff. "Everyone talks about peak oil. But we are at a peak in the consumption of red meat, sugar, furniture and curtains."

Contrary to what most companies do, this company is looking at making its guarantees longer. It wants to ensure its products last longer. In Belgium, this giant company set up a circular economy in furniture. Refresh, revamp, resell, give away or exchange: every option is supported except dumping that old sofa. There are repair workshops, tutorial movies, free spare parts...all this from a company that essentially sells sofas.

The same company also reduced its transport footprint in a radical and pioneering way. Instead of filling boats and trucks with finished furniture that is difficult to stack efficiently, this company decided to leave assembly to the consumer, thus making the transport a good deal more efficient.

So no, multinationals should not be pigeonholed in the same ideologically driven way. The question for me is not how big a business is, but how it operates. The question should also be how governments create market rules where good practices can flourish without the need to compete with those who continue to cut all corners – thereby displacing costs to society and the environment. Governments should not drop to their knees for all multinationals touting jobs, but think about how to create a good environment for companies that do good beyond their shareholders and workers alone. That does not mean that such companies do not have to pay taxes. On the contrary. This means that governments should create a level playing field, not at the lowest common denominator, but at a socially and environmentally sound level.

Yes, the multinational I single out as a good actor exploits all kinds of backdoors to dodge taxes to the tune of billions. But did it lobby for these backdoors or use them to blackmail governments? Or did it simply use the legal tax evasion system that was made by others? Where is the source of that problem?

And yes, this company has factories in China so the assumption that an undercover journalist could find a lot of dirty laundry is not far-fetched. But what counts for this chapter on hope is that positive practices and radical choice editing can and do push the whole market in the right direction.

I get no commission to promote this company, nor do I own shares in it. I have no plans to work for it and hardly ever shop there. To sum up, I have no conflicts of interest. Neither will I make myself popular among that part of the Left that considers all private companies bad by definition. Chances are, you have deduced the multinational in question. But what matters to me is that environmentally positive choice editing and circular economy practices can be found among major companies. However, getting more of them doesn't require patience and a hope for caring CEOs; it requires states to set the rules in such a way that it is these companies that flourish.

Chapter 6

Pioneering movements

Whether it is the Pope, Naomi Klein or Professor Joan Martinez Alier, they all agree that it is necessary to move from the fringe to the mainstream. They all think we should name, question and attack our most important problems. They all identify positive alternatives that should be commonplace but for a lack of political will.

We want politicians to walk the talk, but we can also lead by example ourselves. I happen to like walking in its literal meaning, so much in fact that getting a baby didn't stop my wife and me from completing a 1000-kilometer hike (see my previous book *Walking with Flora*). So in 2014, when I ended a much shorter, maybe 3-kilometer, stroll during Belgium's biggest demonstration against austerity policies in decades, I posted a simple question on Facebook: Who wants to work with me on a multiple day march on Brussels? A march that passes along ecological frontlines in Belgium as well as some great alternatives. A few people responded positively and in January 2015, a group of six came together to start planning for making a wild idea reality: Kris Hofmans, Geert Jespers, Wietse Vermeulen, Jeroen Olyslaegers, Sofie Bouteligier and this author. In March 2015, an adventure started that at least some people would later describe as life changing.

The first 50 participants of the March on Brussels gathered on a pier that extends from the tiny village of Lillo into the River Scheldt. The cold and stormy weather failed to dent an upbeat atmosphere. Greenpeace had arrived much earlier, docking a little boat on the pier and, more remarkably, a tent on top of a massive river-based pylon carrying electricity from Doel nuclear power station to the other side of the river. We shook hands with our more courageous comrades and promised to carry their protest message with us to Brussels, some 70 kilometers walk away. Braving fierce winds laden with toxic air from the massive petrochemical industry in Antwerp's harbor, we arrived at the beautiful domains around Antwerp's North Castle. We met with a group of activist doctors who told us over warm soup that the forest around the castle, tucked between the industry of the harbor and Antwerp city, was a rare green lung, but one destined for the chop: the Flemish government was planning to build a highway extension through it. We convinced a regional television crew to come and ask participants of the march questions – which they did with the celebrities we had managed to bring with us. In the evening we listened to an urbanist who designed an alternative to the current plans, one in which that forest is not only saved but the existing highway would be capped so that a vast swathe of Antwerp is returned to its residents, mostly in the form of a massive park encircling the city. His plan went on to win awards and, after years of struggle, convinced the previously stubborn authorities to go for something close to it. A national nature conservation organization gave us more details on the ecosystems that were at stake in this highway battle. A local social theater company opened their doors to serve the group dinner and another theater company offered a bar and dance floor. With little leg juice remaining and two long marching days to come, most of the pilgrims have by now departed to the dozens of families offering shelter for the night.

The next day, our growing group of hikers meets at a local thrift shop for free coffee and warmth before hitting the road again, striking out towards the CSA farm of Grondsmaak. Along the way, a national level newspaper joins us for a short stretch, giving ample attention to the growing march the next day. Some of the cops walking with us give the children a helping hand by letting them sit on the back of their bikes After almost 30 kilometers we get to Mechelen, a transition town that shelters us for the second evening. Local pioneers share their solutions with the group. On the third day, the still-growing group takes a break at a camp of anarchists who occupy and defend one of the last big pieces of greenery in Brussels, where a new prison is planned. There is a playful action against a proposed shopping village in Machelen and finally we arrive, through the pouring rain, at Allee du Kaai, a Brussels-based co-operative that welcomes us with tasty food made from leftovers and a bar for refugees. Joining the marching celebrities we are greeted by well-known intellectuals and their heart-warming speeches. Everywhere we come, people and associations stand ready to provide the whole group with free food and shelter. A separate smaller group of cyclists that departed at the other end of Belgium also arrive in Brussels that day. Of the hundreds taking part, nobody paid a penny for food or shelter during 3 days. Everything just ran on conviction and sheer goodwill. What started with a Facebook post ended less than half a year later with probably the longest march for the environment Belgium has ever seen.

Our efforts did not happen in splendid isolation. We basically created a wagon that we hooked up to a train that was already going full steam. The next day, we join the Grand Parade of the movement of movements in Belgium: Heart above Hard. This movement organized a huge parade, despite more pouring rain. Over 10,000 people come together to ask for alternatives to divisive neoliberal policies, presenting ten demands that came up through a year of intense discussions among the more active members. But this movement has grown so fast and strong that it's more than just a Parade once a year, or a group of people marching for the environment.

This movement of movements succeeded in keeping the xenophobic Pegida's hate-group from marching in Belgium. Whenever this racist anti-asylum group tried to set up a march, Heart above Hard mobilized such large counter marches that the police had no choice but to call off both. The movement also

Chapter 7

Pioneering governments

A few progressive businesses and movements are essential elements for a great transition – but not enough. Governments need to "see the light" or we will continue heading full speed towards collapse. So the next question is whether there are any hopeful examples of countries or blocks of countries that really dare to address head on the big challenges of our time?

Three examples give me hope, for differing reasons. A Western block of so-called liberal democracies, the most populous country in the world that is officially still Communist and a tiny theocracy tucked away in the folds of the Himalayas. I am referring to the EU, China and Bhutan. None are without fault, but nonetheless all three carry some leadership qualities that the rest of the global pack will need to adopt if we are to get through this century without a major collapse of civilization.

Which political entity does most for the global environment? Until 10 years ago, the answer was pretty easy. The EU began a wave of environmental policies in 1972, at around the time that then commission president Sicco Mansholt finished reading *Limits to Growth* from the Club of Rome. Twenty years later, the EU championed a truly ambitious global environmental agreement at the Earth Summit in Rio de Janeiro. It is from there that, among other things, international climate agreements were born. At home, the bloc has put in place a series of laws and rules that are widely considered the global gold standard for environmental protection. Its Birds and Habitats Directive is one of the strongest nature protection laws in the world. Natura 2000 parks bejewel the continent, sheltering endangered species and inspiring visitors. Energy efficiency laws will soon be saving Europeans nearly €100 billion each year through lower energy bills. Their smart design pushes low quality products off the market while guiding consumers towards the best performing ones, rewarding consumers and innovative manufacturers alike. Within the EU, money circulates from richer to poorer parts, dampening the flames of inequality. It's no longer enough to compensate for the wrongs that make inequality rise, but the principle and structure is in place. To adequately describe the many benefits of the European Convention on Human Rights, we would need a whole separate book. The right to life, freedom, a fair trial, privacy and freedom of speech are just a few indisputable examples. In Belgium, no law can violate the European Convention on Human Rights and I am most thankful we live under its protection.

Does this contradict the many times I have taken aim at European policymakers earlier in this book? Not really. Because it is only in recent years that such figures have been busy trying to dismantle the bloc's achievements. This has happened hand in hand with a weakening of the moral authority of the EU at the negotiation tables where global environmental pacts are made. In the climate summit of Copenhagen in 2009, China sidelined the EU by brokering a deal with India, Russia and Brazil. The US was also spurned, but President Barack Obama was at least in the room when the deal was done. The EU was presented with a take it or leave it deal. Three years later, at the Rio + 20 Summit in 2012, I witnessed the sidelining of the EU first hand.

* * *

After 2 years of hard negotiations over a planetary rescue plan, negotiators were stuck. With just a week left until the summit closed, they were still miles apart on major issues. During a scheduled 2-day break, closed door negotiations continued on a totally new text.

Belgian NGO participants were lucky enough to get access to

the negotiations. In a relatively small room, around 40 negotiators sat around a long U-shaped table. Against the walls, a row of chairs for advisors and a few silent observers, including me. I had scored a front row seat at a geopolitical spectacle. At the head of the table sat the Brazilian chairman. He was brandishing a brand-new text that had been hammered out by his people in secret and in parallel with the formal negotiations over the past 2 years. A stunningly undiplomatic stunt. But the Brazilians also knew that Europe did not want to go home empty-handed. An atmosphere akin to a tropical thunderstorm soon built up in the tiny room. The Brazilian: "We meet here today to agree on the consensus text we gave you last night. We do not have the time to go over it sentence by sentence. It's going to be this or nothing. I urge everyone to adopt a constructive attitude." His finger pierced the air. "I now give the floor to the EU." The EU Representative: "This is impossible for us. If after a 2 years of negotiation of each word, a new text is written, we must at least be able to check every sentence and agree to it, otherwise..." At that very moment, the Brazilian diplomat pushes his microphone button and says: "We will not do that. As I said, we do not have time for that. I give the floor to Korea." The Korean representative: "While we are not completely satisfied with certain pieces of the text and would like to have had more time to discuss, we appreciate Brazil's efforts to reach a consensus. Korea will not veto this text." "Thank you for your constructive attitude, Korea. I give the word back to the EU." "This is not a way of working, we can really..." "We are not going to discuss our method. I'm willing to listen to you if you see something insurmountable in terms of content, but I urge everyone in the room to adopt a constructive attitude. I give the floor to the US." I think I see steam escaping from the ears of the EU representative. He turns around to speak with his two assistants. The former kingmaker of global agreements had practically been ordered to follow in the steps of Brazil or shut up. He and his sidekicks "achieved" by shipping US forests to Europe and burning them. Through my work in the largest federation of environmental NGOs in Europe, I heard horror stories on an almost daily basis from the day Juncker started. Everyone agreed that the attack on European environmental policy was never before so comprehensive.

So that is where we are now: the political dynamic of recent years undercutting exactly that part of policymaking the EU is actually good at. In a recent paper on sustainability, a eurocrat congratulated the EU by looking back at its efforts to make a more egalitarian society based on democracy.[218] Until recently, this might have been well-deserved praise. But Europe's own statistical agency, Eurostat, now shows that inequality has risen over the last decade. And with Juncker at the helm, Europe is also going downhill democratically speaking. Under Barroso, an initiative was launched allowing citizens to push things on the European legislative agenda if they gather a million signatures. All such democratic calls have been rejected or sidelined by the Juncker team. So yes, I have a hard time finding hope in this commission, but not in the EU as an inter-governmental body that can provide great social and environmental protections.

But then again, Juncker's administration is probably only a symptom of a broader trend. In his book *Against Elections*, David Van Reybrouck writes that, "Politics has always been the art of the possible, but today it has become the art of the microscopic. The inability to address structural problems is accompanied by a zealous attention to the trivial, fueled by media that follow the market logic faithfully. The constant stream of futile conflicts is more important than insight into real problems." One only has to look at the UK and the US in 2016 to see to what this can lead. The urgency for the EU to reinvent itself is hard to overstate.

The EU must make a choice: either it will become more social, ecological and democratic or it will continue to fade in popularity and witness the further rise of the authoritarian leaders, fascists and xenophobes that are rearing their ugly heads in several member states. People also need to understand what the EU does for them. It is the EU that brings countries to court because they do not do enough to clean up air pollution. It is the EU that keeps a lot of deadly chemicals far from us. The Juncker team is the dirty bath water, the EU project is the baby. The water must go, but we should be careful to keep the baby.

* * *

Looking for inspiration for a governance system that brings us to a sustainable world for everyone, it makes no sense to cross the Atlantic. Obama had many eloquent words to say about our climate challenge, while at the same time sponsoring fossil fuel projects for totaling $34 billion. That is three times more than former Texan oilman and ex-president George W. Bush. Canada was the first country to rip up its Kyoto obligations. It received a satiric lifetime achievement award from the main global coalition of climate NGOs worldwide for its efforts to screw the climate as hard and fast as possible. Looking instead East, there's something to say about China and it's probably not what you expect.

Jared Diamond is a professor in physiology, geography and environmental science. He is a global authority in the field of the rise and fall of societies. He points to five major factors that contribute to the collapse of a society, with the foremost being environmental destruction. China's fate will hinge on it. Given its size, the fate of humanity as a whole may well be bound to China's.

China is a country with massive environmental problems. It is hard to find a city smog more deadly. Commonplace exposure to dangerous chemicals is mind-boggling. But China is also an island of reforestation on a continent full of countries with rampant deforestation. Every 2 years, China nurtures new forests covering land equal to the size of Belgium.[219] China plants five times more trees than any other country in the world. It was not until 2008 that it created a ministry for the environment, but today that ministry might have more power than the European environment commissioner. China seems to take its environmental problems very seriously.

In the first 4 months of 2015, the country's CO_2 emissions dropped by 5 percent, simply because it was decided that the most polluting coal factories should be closed immediately. If China decides to do something big, it can move fast. Beginning in 2017, owners of coal factories in China woke up with the news that 104 planned coal plants would not be built after all. With 47 under construction, $30 billion of investment was lost.[220] China's leaders decided it was a price worth paying.

Of course, this is not to say all is going great in China. Professor Saskia Sassen documents in her book Expulsions the high numbers of forced displacements and fast-growing inequality.

In China, local corruption or non-implementation of new environmental rules decided at the central level is often the key issue to getting things done. To address that, Beijing is harnessing citizens as a watchdog. Since January 2014, some 15,000 Chinese companies must now publish real-time emissions data, exposing a practice of firing up smokestacks by night. The Party has also encouraged its people to complain to local authorities about maladministration. They can appeal to a new environmental court, which is becoming popular. The Economist wrote about all this: "Although they are a necessity, the top down efforts to strengthen bottom-up pressure may be a signal of a broader trend that will impact far beyond China's polluted air."[221] The Chinese people are venting anger over pollution and the leadership has certainly listened. In recent years, China has become the largest manufacturer and installer of both solar panels and wind turbines, nourished by strong support from government.

While the Chinese state still controls information in a way that is unacceptable to us, it has shown some signs of openness, especially on environmental information. An access to information governance act became effective in 2008 and, by 2011, some 1.3 million requests for information were made. In 70 percent of the cases, the information was given.[222] In many of these cases the information was about pollution, which activists used against companies. In 2015, the penalties for polluters were also increased. Negligent business leaders now risk imprisonment. Protection for whistleblowers was boosted. New penalties came into force for local government officials who do not implement environmental laws. One new law made it so that the penalties for a polluting company increase daily, until the violation of the environmental law has been undone.[223] Within the first 2 months of the new environmental laws, 26 large polluting companies received daily rising fines.[224] Local officials who manipulate pollution figures are being arrested. Another rule ranks local officials on environmental rather than merely economic progress. These examples are part of a centrally organized campaign against the main causes of pollution in China, which is good news for all of us as well. After all, cleaning up the world's factory has environmental benefits far beyond China.

In addition to going after companies and civil servants, the population must also do its bit. Beijing has a new police unit that cracks down on air pollution from relatively minor sources like household barbecues and bonfires. But polluting factories are also temporarily closed on days with high smog. This winter, all non-essential building and demolition projects in Beijing were brought to a halt and the people of Beijing are seeing more and more blue skies again.

Meanwhile, a series of damaging dam projects have also been scrapped in order to better protect the environment. President Xi's concern for the health of the Yangtze River over economic development says a lot about the state's newfound respect for the environment. The grip by the center is tightening. Local environmental protection officials are now centrally appointed. The government has given its Davids a sling with which to attack polluting Goliaths, The Economist writes.[225]

Environment Minister Chen Jining welcomed the launch of an eye-opening film on China's air pollution: *Under the dome.*[226] The movie was viewed by 150 million Chinese in its first 3 days of screening. *Under the dome* did for air pollution in China what *An inconvenient truth* did for climate change in the US: catapult the issue to the top of public and policy debates. Only after 300 million Chinese people had watched the movie and the public anger started to erupt, did Chinese censors intervene. So yes, there is still a long way to go in terms of freedom of the press and expression. But at the same time, there is clearly some opening up when it comes to environmental issues. Increasingly, to climb the party ladder, you need to work on your green credentials.

To sum up: China is busy containing its largest environmental offenders, throwing bosses in jail if they knowingly break pollution laws, raising fines on polluting companies, helping citizens to access information and environmental justice, and assessing local authorities on their ecological achievements. The environmental lawyer James Thornton, famous for never losing a case in 4 decades of practice, says China's "ecological civilization" concept is the best response to the world's environmental crises. China has even changed its constitution to get that in. For all the flaws in the Chinese governance model as a whole, it still seems to me that we need to look beyond our differences and learn a thing or two about how the Chinese are dealing with their massive environmental issues.

* * *

Neighboring China, but tucked away in the folds of the

Himalayas, is a small theocracy with a growing democratic accent. Bhutan is a small Buddhist kingdom of less than a million inhabitants and a cool dragon emblem on its flag. But the most interesting thing about Bhutan is that the country has no aims to grow its economy. It chooses instead to aim for an annual increase in gross national happiness (GNH). And yes, this encompasses a whole lot more than some Buddhist ideas about increasing spiritual rather than material wealth. In many ways, Bhutan shows us the way to a better future.

What started with a statement from the former king in the peace and love era of the 1970s has evolved into a policy with four pillars: sustainable socio-economic development, environmental conservation, preservation and promotion of culture, and good governance. These laudable principles were enshrined into its constitution and, from 2008 onwards, there has been a GNH index, the guiding light for all national policies.

The Bhutanese go for progress on matters of psychological well-being, health, time use, education, cultural diversity and resilience, good governance, community vitality, ecological diversity and resilience, and living standards. They use 124 indicators to measure progress. The Bhutanese definition of happiness goes beyond the superficial and temporary "feeling good". According to the government, happiness comes from living in harmony with nature, with other people and with Bhutan's heritage. Being connected to the world that surrounds each human is what human development is all about, for people in Bhutan at least. Putting everything in function of GNH has far-reaching consequences. After a test with the GNH indicators, Bhutan decided not to become a member of the World Trade Organization. It also decided that large mining operations would be bad for GNH. It keeps mass tourism away by charging stiff entry fees and allowing only organized travel with a state guide. The result of doing so many things differently? The World Bank calls Bhutan a champion in the reduction of poverty. Universal basic education is a fact. In 22 years, life expectancy increased by a whopping 16 years. Plastic bags are forbidden. Bhutan takes more carbon out of the air than it adds to it, probably the only country in the world to do this. It is stated in the Constitution that 60 percent of the country's forest must always remain and Bhutan is shifting to 100 percent organic farming. These are policy results that many people all over the world would love to sign up for. If you compare Bhutan with nearby Nepal, the difference can hardly be any bigger. Until 1950, Nepal was more closed than North Korea is today. But once the gates were opened, it became a hippie hotspot and a donor darling. Capitalism was added to a feudal caste system and the results speak for themselves. Today, Nepal is a whole lot poorer than Bhutan and is one of the poorest countries outside Africa. After half a century of going along with goals set for it by donors, most of them Western, it still scores badly in areas from education to health and from nature conservation to harmony. How the world interfered with Nepal and what effects this has for ordinary Nepalese was the subject of my earlier book Nepal. Nieuwe wegen in de Himalaya, based on 2 years of fieldwork in the country.[227]

But Bhutan is not a paradise on earth either, of course. Some people are still unhappy, like the many Bhutanese of Nepali origin who were kicked out after being in Bhutan for several generations. There are also poor people in the country and yes, democratic participation remains minimal. Nevertheless, a form of governance that makes so much social-ecological progress for so many people in such a short time should inspire us. The Bhutanese government also wants to export the GNH model. Bhutanese policymakers realize that Bhutan too will go down if it remains an island in a neoliberal ocean. "We know that our glaciers melt due to the emissions of cars in cities like Los Angeles," said former Prime Minister Thinley. In April 2012, Bhutan chaired a conference on happiness and well-being within the UN. It prompted an international panel of experts to write a UN report under the guidance of Bhutan: *Happiness: Towards a New Development Paradigm.*[228] They identified two basic principles at the heart of this new paradigm: the universal pursuit of happiness and the existence of planetary boundaries. The authors argue that the doctrines of endless growth and the pursuit of a status through ever increasing consumption are untenable. By the way, between 2001 and 2011, Bhutan's GDP increased by 8.4 per cent a year. By accident. This put it in the illustrious top ten of best performing countries on the scoreboard of neoliberal economists. It was not a goal, it just happened en route to becoming a better place to live. And to be clear: for a country that had such a low GDP in 2001, GDP growth was a welcome *by-product*.

A different, peer reviewed publication charting Bhutan's success ends with a disturbing finding.[229] According to author Anders Hayden, the strength with which Bhutan is maintaining an alternative economic model is decreasing. Pressure on the model is coming from three sides: the growing private sector, the penetration of consumer culture through TV and the internet, and the growing democratization of Bhutan. The latter "threat" in particular raises a tough question: is parliamentary democracy, as we know today in countries in the West, suited to the many challenges outlined in this book? David Van Reybrouck is convinced that parliamentary democracy is becoming less successful. He says that each management system stands or falls with efficiency and legitimacy. On both fronts we are losing traction. Throughout the Western world, the past half century has seen ever fewer people voting during both national and European elections. It also takes longer to form governments. The power that our parliamentary democracies have to deal with the really big issues is like the ice extent at the North Pole: it fluctuates with the seasons but the general trend is clearly going down pretty fast.

The famed Harvard University is fortunate enough to have

Yascha Mounk teaching its students about the decay of Western democracy in the last century. In January 2017, he published an important scientific article in the Journal of Democracy.[230] Almost 75 percent of all people born in the West in the 1930s claim that democracy is essential. With people born in the 1980s, this opinion is shared by only around 25 percent. The younger the interviewee, the more support they gave an authoritarian and military regime. The author has embarked on fresh research, collecting examples from reactionary regimes from Poland under the PiS party to Donald Trump's America. Mounk is not even sure that democracy will survive Trump.[231] The Donald himself indicated he would only respect democracy if he won. British political scientist David Runciman is another skeptic. He thinks we are naive to believe that because democracy has survived the crises of the past decades, it will do so in the next few. As Trump's plans are likely to make the poor and frustrated even poorer and more frustrated, it is likely that the trust in democracy will diminish further. Already, one in six in the US now think that the army should be in charge. That belief is also rising in Europe.

Maybe the election of Trump is the wakeup call that will reverse this trend. After all, his inauguration was followed the next day by the biggest demonstration ever to take place in the US, one summed up by the word "resist". Some pundits, activists and political scientists think that something good can come from the unity that comes with the opposition to Trump. Left-wing philosopher Slavoj Zizek feels that only Trump could stimulate such an awakening. But then what? I mean, if people are awake enough to see that "democracy" has produced President Donald Trump, then what? David Van Reybrouck is one of the few with an answer. He has studied a principle common to democracies from ancient Athens to the French Revolution: the lottery. The idea is simple. Select a representative section of the population by lottery and put them together. Give them round tables, snacks, tea, coffee and a few independent technical experts to run ideas past. Do this a couple of times over the timespan of a year and you end up with better policies than the current *partycracy* is coming up with. This is not as wild as it first may seem. In fact, Belgium has such a system when it comes to life or death decisions. A jury is a random selection of upstanding citizens tasked with reaching a verdict on the facts established by criminal courts. And then there is the G1000 that Van Reybrouck himself organized. A group of 1000 people who pondered rather long and hard about 25 major issues of our time and came up with concrete recommendations. The initiative was repeated in a series of cities in the Netherlands. But it has not been recognized yet as a new form of democracy, one that might be better than what we have today, which is in theory a parliamentary democracy, but in practice a party-political system. But the signs that this might be changing are cropping up beyond Western Europe as well. Since 2006, pirate parties have hoisted sail throughout the Western world, hell bent on shaking up the way power is exercised. They came closest to seizing the helm in Iceland. Their battle cries called for direct democracy, citizen participation in all major decisions, full transparency and freedom to exchange ideas. After The Guardian newspaper's Panama Papers exposed wrongdoing in the Icelandic government, the party surged in the polls to over 40 percent. But the winds dropped from its sails somewhat on election night and it took "only" 15 percent of the popular vote. In 40 other countries black flags are clearly visible in national politics, but only on the horizon. The fact they are visible at all gives a sure signal: our democracies are in need of a total overhaul. In Europe there is also a very European party: DiEM25. It wants to write a new European Constitution, by citizen groups and a truly democratic EU. Vested interests are of course doing what they can to stop their own boats from rocking.

Nowadays, the biggest differences are not between a social liberal, a liberal socialist and a conservative democrat. A growing mass of people are tossing them all onto the dustbin of history. Old mainstream parties have melted faster than the ice in the Arctic, especially Europe's traditional socialist parties that have gradually abandoned their founding values. Today, the major political frontline is between authoritarian and often also racist leaders and deeply democratic progressives. It is a battle between Victor Orban and Yanis Varoufakis, Geert Wilders and Jesse Klaver, Marine Le Pen and Pablo Iglesias, Donald Trump and Bernie Sanders, Nigel Farage and Jeremy Corbyn. All have had momentum, but when it comes to actually capturing power, it has until now been mostly the far-right who have won the battle of the exit-doors. But that's the state of play in autumn of 2018. By the end of 2020 a lot of cards might have changed hands from the keepers of the authoritarian right-wing exit door to those opening the door to a deeply democratic left.

For more than 200 years, we have been working hard to improve the system of wealth redistribution that ensures care for our young, sick, our homeless and elderly. But globalization and automation have been reshaping jobs and the economy, while our redistribution model remains based on labor tax incomes. This is causing it to fall apart. One approach to address rising inequality could be to cap wages while also making a rule that the highest wage can never be more than ten times the lowest wage. There are also so many other ways to reduce the expenses of the state than taking from the weak. No law of nature says that Belgium should give a €4 billion fossil fuel subsidy to cars, which it happens to do. The key obstacle to both solutions is an unshakable belief in the free market. Its backers in the political elite do not want to price in environmental and other harms to products. Taxes on things like diesel are a limp fig leaf. Green from afar, very thin and not at all able to hide the ugly truth.

The good news is that we are finally ready to have a debate on the big myths of our times. That debate is urgent and necessary. More and more people are finding their way to a

Conclusion

A recent essay in a Flemish quality newspaper carried a title that struck me: "The truth exists (but doesn't matter)", Subtitle: "Why facts are not so important in politics."[232] That's scary, at a time when 20,000 scientists say we're fucked if we don't change our ways. At the end of 2018, even the too soft and slow IPCC panel said that "unprecedented changes are needed"[233].

If facts would be important in politics, all politicians would say that the era of oil, coal, gas and uranium is basically over, and having gorged on it, so is the relatively stable climate in which humanity has thrived. They would say: "our economy is built on sand, but the sand is running out of stock". They would admit that until now, they have utterly failed in addressing the sixth mass extinction of all life on earth and tell all voters that no matter what they once promised, this civilizational crisis is now too big to put behind other priorities. What humanity needs is political leadership that declares war on mass extraction to avoid mass extinction. What humanity has is a liar-in-chief at the White House. Trump promotes the extractive industry like no other, because he doesn't care about facts and because the scene has been set for fact-free politics over the years.

The truth is constantly under siege from a polluter's paid army of merchants of doubt. They sell lies for a living. Too many corporate-funded media give unchallenged airtime to such truth-deniers. We need the most inconvenient facts of all always at arm's length. When I got a chance to speak at a business TV channel I ignored communication rules (and the proposed frame) to talk about the sixth mass extinction of life on earth, because I felt it was necessary that the science on this also reached an audience that is used to discussing barriers to more economic growth.

Some people confuse my message with my personality, or say

they get depressed from my book. That's neither the purpose nor helpful. It's not because there's a lot of bad news that I'm a negative person, pre-coffee morning's aside. I do believe we can change the way humanity lives on this earth, radically. Private banks can be public, or in the hands of groups of people with good ideals on what to fund and what not. Similar story for energy, water, schools and more. There's no nature law forcing us to trade in hot air to enrich traders and industrialists while not helping to reduce emissions. We can quit calling everything that is of value a "trade barrier". It all starts by admitting that when it comes to trying to save humanity from an ugly civilizational collapse, greening capitalism failed as a solution and is in fact a dangerous illusion. The choice we face is between sleepwalking to a civilizational crash that will probably be unlike anything since the fall of the Roman Empire or to reconnect the bonds in communities and with the earth and from there find a social and fair way to radically slow down a vastly overheated industrial economy before it burns out all people. Most green labels and green growth myths are only *hopium* for the people. False hope to keep us from coming together and revolting.

* * *

Facts are important, but only stories stir people into a revolution. Stories have far greater power to move people than facts and that's why I choose to bring story after story to show what is going on. Laudato Si from Pope Francis was just an example of a positive religious rewriting of the greatest of all stories, the story of the journey of humankind of planet earth. I've been particularly inspired and moved into action by several touching stories of people on the frontlines, which is one reason why I wrote this book. The better story doesn't have to be confined to one particular -ism but it sure can't be a form of neoliberalism or capitalism. Neoliberalism is just a way to put capitalism on steroids. Capitalism needs growth but the earth's resources aren't giving us that option. Capitalist growth is causing ever more conflicts to pop up all over the world. The +-3000 conflicts in the Atlas of Environmental Justice are like the always harder hurricanes and hotter heatwaves: warning signs of a system that is failing. Naomi Klein was right when she said it's capitalism versus climate, but not complete. The climate breakdown is just one of so many signals that illustrate how humans have overreached. All the evidence is in place to say that it's capitalism versus thriving as a species. Any big story about the future journey of humanity on planet earth that does not put into question the volume of stuff we put into the global economy or the rising costs of pursuing GDP growth is a fairy tale, a distraction. In short: the new story will have to be a positive post-growth story.

* * *

As a European I can't conclude without saying something that strikes me about the risks facing Europeans in particular. I simply can't escape the impression that the European project is a lot more fragile than it seems from the surface of it. I'm not talking about Brexit, which, at least for the rest of the EU, is just a nasty bump in the road. When I see the European project, I see exponential volumes of mostly unfair trade imports and I wonder what will happen when the competition for the world's dwindling resources gets even more brutal and people all over the world refuse to play the neoliberal free trade game that the EU wants it to play. The EU has already snatched an unfairly big slice of the cake and all over the world, people and their governments are sick of it. Not only has all the low hanging fruit been picked already, the EU is increasingly arrogant by thinking it can simply continue to pick from other more remote orchards – as it does with everything from uranium to wood. But it all requires ever more resources, coercion and violence. Our hands are already black and dripping with oil from tar sands in Canada. The fuel only barely warrants the name, consuming almost as much energy as it gives and leaving behind what could easily pass for Mordor in Tolkien's book Lord of the Rings. Our oil imports are the reason for supporting so many brutal regimes. For gas we "solved" our unhealthy dependency on an anti-democrat East of Europe by adding an unhealthy dependency on an anti-democrat West of Europe, across the Atlantic. But it's not just our unhealthy fossil fuel addiction and how Europe's dependency on it is detrimental to democracies in and out of the EU. We're chopping wood all over the world like no other, with subsidies whose official use is preventing climate change. We send our e-waste and toxic ships in return. That won't last. China's ban on importing our plastic waste and Trump's trade tariffs are early warning shots. It is the stubbornness of a neoliberal European elite that puts Europeans in this fragile position while paving the way for the extreme-right to steer a justified frustration of Europeans towards some easy scapegoats. Where are the other European leaders who are not from the far right with the balls to say that what Europe really needs is a well planned and executed de-globalization, for all the good social and ecological reasons? We need protectionism. Not patriotic protectionism but ecological protectionism: a trade policy that faces the ugly truths about the century we live in.

* * *

Of course it's not just the EU that has an addiction problem from which a rude awakening is bound to happen. The whole GDP focused global economy is akin to a desperate junkie, hooked on GDP. Either by its own will or with a little help from the IMF dealer. It gets its kicks from fossil fuels and it leaves its dirty needles lying around. As an example: every year, oil leaks in the

Niger Delta are equal to that spilled in BP's massive Deepwater Horizon disaster. The GDP junkie badly needs rehab, but he's far beyond reason or arguments to go there. Who's going to bring him in?

It turns out there are ever more volunteers for making that happen. One group in particular is boxing far above its weight. Indigenous peoples make up only a few percent of the population worldwide, but in almost half of all environmental conflicts, they man the battle line. It sounds simple but it has strong empirical backing; if we protect them, our frontline defenders, we protect our common home, the earth. The number of them being killed has risen fast, from one to four martyrs a week. It's about time we recognize them as martyrs not for their community, a country or God but for all of us on this earth. Facts and stories both tell us to protect the protectors. We need them as much as they need us.

But this resistance goes far beyond the bravery of so many indigenous peoples. There are always precedents but it's fair to say that in this global war, a new round of global resistance began at the turn of the Millennium with the successful undermining of the World Trade Organization meeting in Seattle. Later, movements like Occupy and Anonymous followed. Now there is Blockadia, the collection of direct actions trying to stop environmental destruction. There is a financial divestment movement, draining the bad guys of their finance mojo. There is the degrowth or postgrowth movement that reveals our most fundamental myth, GDP growth, as one big lie. This movement is busy spilling from the activist and academic spheres into the political sphere, with members of the European Parliament coming straight from that movement and a postgrowth conference in the European Parliament organized by members of five different European party groupings, as well as by this author. Then there are transition movements that show how things can be done very differently at the local level. There are civil movements like Hart boven Hard, which combine local actions with national alternatives to the austerity regime. The divestment movement convinced around a 1000 institutions to ditch around €7000 billion in fossil energy industry stock. The largest private coal company in the world went belly up in 2016, as did a record number of oil and gas companies.[234] At the time of writing these last lines, just before the 2018 US mid-terms, signs of a political swing in the US were on the rise and the model that got Bernie Sanders so far is bringing justice democrats into Congress. Imagine what could be possible if in 2020 in the US, an eco-socialist like Bernie Sanders rose out of Trump's ashes, or what a revolution it would be if Jeremy Corbyn took the helm in the UK. Neoliberalism is finally dying and that's a good thing. Yes, the bad news is that fascists are leading the new narrative but this could still flip in a radically different direction, if we open enough eyes before someone shuts them down the hard way.

Participating in the global resistance and protection movement is a double-edged sword. First there's the living differently thingie. Eat, live, move, clothe, pay and do almost everything greener. In my view, finding fun and low footprint lifestyle hacks just gives positive energy for yourself and it usually connects you with other people. But I fully realize we are never going to save humanity from a major collapse by just urging everyone to cycle. Eating tofu won't stop Chevron. This book is not about the individual choices you and I make, but about our collective choices. Personal change merely gives me the positive energy to deal with the more difficult part of the struggle, the political battle against injustice and degradation on a global scale. Even the greatest transition town is destroyed when a category 5 hurricane hits. Buying an electric car will not stop AREVA from ruining Niger to supply us with nuclear power, rather to the contrary.

The revolution must be collective and it must also be social and ecological. Even that will not be enough. We have to fix our broken democratic system. The revolution will need to overcome over 2 centuries of capitalism. Does that mean throwing everything we know out of the window and every culprit under the bus? Well, not if you ask me. Despite all the critiques that can be made on them, I think that we can use many good EU laws and even the way they are made, with a thriving civil society as an important player. I also believe in Chinese know-how on how to be bold and also implement anti-pollution laws. From Bhutan we can sure learn something about a holistic vision that comes with better targets and indicators than GDP. I'm against religious politics but pragmatic enough to advise reading and spreading one very useful text from the Pope. In all examples, as well as in the example of the multinational with at least a few really good practices like radical choice editing and industrial upcycling, I think we need to be pragmatic as well as revolutionary. We should park prejudices, air well-deserved critiques where appropriate but still look at everything that is useful with a fresh mind. As I wrote earlier: some schizophrenia will be needed. Some mental flexibility on the set of ideas that can help us forward will be needed. I hope we can cherish a *pluriverse* of narratives, focus on what works and just do things. Get the ball rolling.

* * *

The pioneers highlighted in this book have often been forced to choose sides. Most have been angry, sad and fearful at some point in their struggles. All faced seriously polarized situations that were created not by their choice. Rather than being consumed by their feelings, they turned them into constructive action. Their gaze remained steady, however uncomfortable the truth was to behold. Easy but compromised ways out were always there, but they recognized them for the dead ends or flimsy excuses they are. They loved others and our earth as much as themselves. They also found connections with other heroes on other frontlines far away. They formed their multinationals, those of the resistance. That gives me some reason for hope at a time when the relationship between humanity and ecosystems on earth is in an undeniable crisis. Frontlines have opened everywhere. On the one hand, you have major mining companies, neoliberal economists, post-truth politicians, merchants of doubt and a small but powerful elite that has lost all sense of fair play and justice. Rising from the swamp they made in the first place are fascists and racists, who sharpen their daggers against refugees and other scapegoats. But on the opposite side are people who stand for fairness and who see through all that. Groups that want to close coal mines instead of borders. People with due respect for the planet we live on. They don't just want the polluters to pay for cleaning up the crisis they have caused, they want an end to the "co_2lonisation" of the earth. More and more people in this group are either forced to find themselves on a frontline or volunteer to be there. Why not join them voluntarily before some frontline opens up at your doorstep?

The final question for you, dear reader, is not so much which party will win this political, ecological, economic and global struggle that is raging around us. Nobody can give you any guarantees on that. The key question is: what will you do after putting down this book?

Note to reader

From Nick Meynen: Thank you for purchasing Frontlines: Stories of Global Environmental Justice. My sincere hope is that you derived as much from reading this book as I have in creating it. If you have a few moments, please feel free to add your review of the book to your favorite online site for feedback. Also, if you would like to connect with other books that I have coming in the near future, please visit my website for news on upcoming works, recent blog posts and to sign up for my newsletter: https://nickmeynen.wordpress.com/.

Email: meynen.nick@gmail.com

Facebook (NL): https://www.facebook.com/nick.meynen

Twitter (EN): https://twitter.com/NickMeynen

Sincerely, Nick Meynen

Acknowledgments

On my journey I was lucky enough to receive help and inspiration from a large number of people. While I can never do justice to all of them, I want to thank some personally.

When Leida Rijnhout hired me in 2008 and placed me in an international project with Professor Joan Martinez Alier, it felt like finding my mission at the age of 28. I see them as my mentors, a Godmother and Godfather. I'm grateful that they've been supporting me throughout my learning curve.

Since joining the European Environmental Bureau (EEB) in 2014 I have also greatly benefited from the specialist knowledge of many expert colleagues on various environmental matters. I especially thank Jeremy Wates for seeing value in the difficult systemic change debates that I stirred from time to time within the EEB, and for allowing me to create a new position: policy officer on environmental and economic justice.

But here's a disclaimer: nothing in this book is the opinion of the EEB or its millions of members all over Europe. This book is neither written during my working time nor financed by the EEB and it should in no way be considered as an EEB product.

I like to thank writers, journalists and mentors Gie Goris and Isabelle Rossaert for opening their homes for inspiring retreats. Excellent feedback on draft manuscripts was given by Marjan Cauwenberg, Wietse Vermeulen, Johan Denis and Hans Devroe. For the nitty gritty of spelling, grammar and source checks, Mercedes De Grande, Jan Drieghe and Inge Verbeeck offered me a great service.

And then there are the many people who helped me during fieldwork. For the chapter on Greek gold alone, I wish to thank Stavroula Poulimeni, Maria Kadoglou and Charlotte Christiaens: they were indispensable. My toxic tours in Italy and Bulgaria were only possible thanks to Lucie Greyl and Evgenia

Tasheva, while I have Steven Spittaels to thank for giving me the opportunity to map the war in Eastern Congo.

A big thanks to all the people I had multiple or longer interviews with, in particular to Bruno Chareyron, Georgi Kotev, Katrien Van der Biest, Sumaira Abdulali, Julio Prieto, Steven Donziger, Alexandru Popescu, Nic Balthazar, Roger Cox, Jan van de Venis, Filip De Bodt, David Dene, Tom Troonbeeckx and Patrizia Heidegger.

The chapter "The trade in hot air" would not exist without the financial support of the Pascal Decroos Fund for investigative journalism: a fund that gives journalists the chance to dig much deeper than usual. The chapter "Sustainable destruction" was only possible with the help from the organization Medicine for the People and their dedicated staff in Hoboken. I'm very grateful for the funding for translation received from the Institute of Environmental Science and Technology - Autonomous University Barcelona (ICTA-UAB), Project: EnvJustice, European Research Council (ERC) Advanced Grant 2016-21. However, I'm the sole responsible person for the content. Neither ICTA-UAB nor the ERC can be held accountable for the contents of this work.

Both the former and the current publishers at EPO deserve great credit for not just accepting my original book but for their intense guidance towards the finished product in Dutch. In the journey from the Dutch original to this book I've had great support from English native speaking colleagues Anton Lazarus and Anita Willcox, who often polished a pitch or synopsis in exchange for as little as a homemade rhubarb pie. Of a different category was the year-long support from Jack Hunter, who reviewed the whole manuscript line by line, after my rough translation. But that was worth every penny Jack, as you also translated my wordplay, humor and nuances to do justice to my writing style in Dutch.

"I just need to finish this sentence" is a line I can no longer utter at home, after years of stretching the limits of buying time to work on my book. Fany Crevecoeur and Flora Meynen-Crevecoeur deserve apologies, compensation and a very big thank you for all those days and evenings that they tolerated me being home but not really being with them. I want to thank them for the understanding, patience and for the moral support. Our youngest family member, Rosalie Meynen-Crevecoeur, was too young for all of that. But I also took dad-time away from here to write this book. I've often had her in mind when writing this book, as she has a life expectancy that stretches into the twenty-second century. If she makes it to that century, what will she see in the review mirror? Will she feel just as thankful as I feel towards all those heroes that feature in this book?

References

1   Temper, L., 'Niyamgiri-Vedanta Bauxite Mining, India', 9 November 2014, https://ejatlas.org/conflict/niyamgiri-vedanta-bauxite-mining-india

2   Foil Vedanta, 'Protest at Vedanta HQ as Supreme Court decision announced', http://www.foilvedanta.org/news/protest-at-vedanta-hq-as-supreme-court-decision-announced/

3   Shinde, M., Bokil, A., '13 protesters against copper plant in India killed after police open fire', The Ecologist, 25 May 208, https://theecologist.org/2018/may/25/13-protesters-against-copper-plant-india-killed-after-police-open-fire

4   Environmental Justice Atlas (https://ejatlas.org/)

5   Meynen, N., 'Nucleair ramptoerisme in Bulgarije', in *MO* Magazine*, September 2011 and Meynen, N., 'Uranium from Russia, with love', in The Ecologist, 4 August 2016, https://theecologist.org/2016/aug/04/uranium-russia-love

6   David Suzuki Foundation, 'The Water we drink', p.18, November 2006, http://www.davidsuzuki.org/publications/downloads/2006/DSF-HEHC-water-web.pdf

7   Wikipedia, 'Summers memo', https://en.wikipedia.org/wiki/Summers_memo

8   CRIIRAD, http://www.criirad.org/

9   http://www.nuclear-free-future.com/en/laureates/laureates/bruno-chareyron/

10 Meynen, N., 'France destroys North Niger. Will the EU or UN act?', 11 October 2017, https://www.mo.be/en/opinie/france-destroys-north-niger-will-eu-or-un-act

11 Meynen, N., Poulimeni, S., 'Greek goldrush shows that Panamapapers are tip of the iceberg', 20 April 2016, https://www.mo.be/en/analysis/greek-goldrush-shows-panamapapers-are-tip-iceberg and Meynen, N., Poulimeni, S., 'The Greek state has nothing to gain but environmental

cost from the investment', 21 April 2016, https://www.mo.be/en/analysis/greek-state-has-nothing-gain-environmental-cost-investment

12   Kentikelenis, A., Karanikolos, M., Reeves, A., McKee, M., Stucker, D., 'Greece's health crisis: from austerity to denialism', in *The Lancet*, 2014 http://www.enetenglish.gr/resources/article-files/piis0140673613622916.pdf

13   Papazachos, K., 'Seismological advice on the tailing dam in Skouries' (free translation from the Greek document), http://bit.ly/2lArJpW

14   Ciobanu, C., 'Roemeens parlement stemt tegen carte blanche voor mijnbouw', op www.mo.be, 11 December 2013, http://www.mo.be/artikel/roemeens-parlement-stemt-tegen-carte-blache-voor-mijnbouw

15   World Wildlife Fund Greece, 'Report on violations of EU law in the watershed management plan for the central waterdistrict of Macedonia' (free translation from Greek), January 2015, http://www.wwf.gr/ images/pdfs/WWF-Report-to-European-Commission-Skouries.pdf

16   International Monetary Fund, IMF Country Report nr.13/241, July 2013, http://www.imf.org/external/pubs/ft/scr/2013/cr13241.pdf

17   Van Os, R., McGauran, K., Römgens, I., Hartlief, I., 'Fool's Gold', March 2015, SOMO, https://www.somo.nl/fools-gold/

18   http://www.ekathimerini.com/168985/article/ekathimerini/comment/mining-the-depths-of-frustration-in-halkidiki and http://www.kathimerini.gr/810291/article/proswpa/syn entey3eis/pol-rait-an-synexis8ei-h-katastash-stamatame

19   Römgens, I., 'The "dark side" of the Netherlands', 19 May 2014, SOMO, https://www.somo.nl/the-dark-side-of-the-netherlands/

20   Smith H., 'Greece retries journalist who leaked "Lagarde list" of suspected tax evaders', in *The Guardian*, 8 October

2013, https://www.theguardian.com/world/2013/oct/08/kos tas-vaxevanis-greece-lagarde-list-tax-evaders

21 https://wikileaks.org/imf-internal-20160319/transcript/ IMF%20Anticipates%20Greek%20Disaster.pdf

22 ASADHO, CDH, CVDHO, 'Nord-Katanga. Attaques délibérées contre la population civile', October 2003, https:// www.ecoi.net/file_upload/bsvec1_Rap_final_Nord_Kat. pdf

23 Meynen, N., Spittaels, S., 'Mapping interests in conflict areas: Katanga', August 2007, IPIS, http://ipisresearch.be/ publication/mapping-interests-conflict-areas-katanga/?_sf_ s=Mapping+interests+in+conflict+areas:+Katanga

24 Vandaele, J., Goris, G., 'De lange tenen van George Forest', 16 May 2008, http://www.mo.be/artikel/de-lange-tenen-van-george-forrest

25 Kaplan, R., 'The Coming Anarchy. How scarcity, crime, overpopulation, tribalism, and disease are rapidly destroying the social fabric of our planet', February 1994, *The Atlantic*, http://www.theatlantic.com/magazine/ archive/1994/02/the-coming-anarchy/304670/

26 Wikipedia, 'Development-induced Displacement', https:// en.wikipedia.org/wiki/Development-induced_displacem ent#cite_note-2

27 Piketty, T., *Capital in the Twenty-First Century*, Harvard University Press, 2014

28 Economic Policy Institute, http://stateofworkingamerica. org/chart/swa-wages-figure-4-ceo-worker-compensation/

29 Hanauer, N., 'The Pitchforks Are Coming...For Us Plutocrats', July/Augustus 2014, *Politico Magazine*, http:// www.politico.com/magazine/story/2014/06/the-pitchforks-are-coming-for-us-plutocrats-108014

30 Holslag, J., 'Wees er maar zeker van: die middenklasse, zij gaat eraan. Ik gruw van deze globalisering', 26 October 2016, in *De Morgen*, http://www. demorgen.be/opinie/

wees-er-maar-zeker-van-die-middenklasse-zij-gaat-eraan-b9da2cef/

31 Vandaele, J., 'De middenklasse krimpt, het werk precariseert. Leve de 21ste eeuw?', 3 June 2016, http://www.mo.be/analyse/de-middenklasse-krimpt-het-werk-precariseert-leve-de-21ste-eeuw

32 UNEP, 'Sand, rarer than one thinks', March 2014, https://na.unep.net/geas/getUNEPPageWithArticleIDScript.php?article_id=110

33 http://gulfnews.com/business/oil-in-dubai-history-timeline-1.578333

34 https://aleklett.wordpress.com/2014/10/31/dubai-as-oil-producer/

35 http://www.bbc.com/news/business-16494013

36 *Dredging Today*, 'Malaysian Sand Issues', August 2010, http://www.dred-gingtoday.com/2010/08/05/malaysian-sand-issues/

37 *Global Witness*, 'Shifting Sand', 10 May 2010, https://www.globalwitness.org/en/reports/shifting-sand/

38 http://mothership.sg/2017/04/sand-exported-unethically-from-vietnam-end-up-in-singapore/

39 http://foreignpolicy.com/2010/08/04/the-sand-smugglers/

40 http://www.smh.com.au/news/world/singaporeindonesiaborderdisputebuiltonsand/2007/02/16/ 1171405445847.html

41 https://www.wired.com/2015/03/illegal-sand-mining/

42 https://www.washingtonpost.com/world/sumaira-abdulali-fights-to-lower-noise-levels-in-mumbai-indias-capital-of-noise/2013/10/02/bf9e2c8c-26b5-11e3-9372-92606241ae9c_story.html?utm_term=.09b1960c75b8

43 http://mumbaimirror.indiatimes.com/mumbai/other//articleshow/15889768.cms?

44 http://timesofindia.indiatimes.com/city/mumbai/HC-bans-sand-mining-across-Maharashtra/articleshow/6623432.cms

45 https://www.youtube.com/watch?v=sZNB9GtixXo

46 http://www.thehindu.com/news/cities/mumbai/news/
Vaitarna-rail-bridgeís-foundation-damaged-by-sand-
mining-WR/article14552228.ece

47 http://www.thehindu.com/news/cities/mumbai/news/
Vaitarna-rail-bridgeís-foundation-damaged-by-sand-
mining-WR/article14552228.ece

48 http://www.thehindu.com/news/national/other-
states/Mahad-bridge-collapse-10-bodies-recovered/
article14551026.ece

49 http://www.dnaindia.com/india/report-sand-mafia-free-
run-abetted-mahad-tragedy-say-greens-2241220

50 https://thewire.in/tag/beach-sand-mining/

51 https://scroll.in/article/832422/for-a-month-now-a-chennai-
reporter-who-exposed-illegal-sand-mining-has-been-
receiving-threats

52 https://thewire.in/116202/sand-mafia-expose-harassment-
intimidation/

53 http://www.ejolt.org/wordpress/wp-content/uploads/201
5/02/FS_010_India-Sand-Mining.pdf

54 http://suchetadalal.com/?id=cf00fa41-5153-e2c0-
4ba232f95eee&base=sections&f

55 http://suchetadalal.com/?id=cf00fa41-5153-e2c0-
4ba232f95eee&base=sections&f

56 http://www.thebetterindia.com/43685/plastic-waste-in-
road-construction-plastic-man-india-prof-vasudevan/

57 http://www.dbexportbeer.co.nz/db-beer-bottle-sand

58 http://ejatlas.org/conflict/mechanized-sand-mining-in-the-
maha-oya-sri-lanka

59 Lyme, C., 'Sand winning: Ghana's Disappearing beaches',
video on Vimeo, https://vimeo.com/72047697

60 Van Lancker, V., Lauwaert, B., De Mol, L., Vandenreyken,
H., De Backer, A., Pirlet, H., 'Zand- en grindwinning',
2015, In: Pirlet, H., Verleye, T., Lescrauwaet, A.K., Mees, J.
(red.), *Compendium voor Kust en Zee 2015: Een geïntegreerd*

kennisdocument over de socio-economische, ecologische en institutio- nele aspecten van de kust en zee in Vlaanderen en België. Oostende, Belgium, p.109-118, http://bit.ly/2lGFFzx

61 De Walsche, A., 'Jan De Nul: voorbij de verontwaardiging', 26 June 2013, http://www.mo.be/artikel/jan-de-nul-voorbij-de-verontwaardiging

62 Verwaest, T., 'De impact van aggregaatextractie op de kustveiligheid bij storm', 2008, Waterbouwkundig Laboratorium, http://bit.ly/2lGnMAQ

63 Vlaamse Bouwheer, 'Metropolitaan Kustlandschap 2100', December 2014, http://doc.ruimtevlaanderen.be/rapport/20150226-MKL2100-Eindrapport-Deel1_.pdf

64 Wikipedia, 'Precautionary Principle', https://en.wikipedia.org/wiki/Precautionary_principle.

65 Van Lancker, V., Lauwaert, B., De Mol, L., Vandenreyken, H., De Backer, A., Pirlet, H., *idem*

66 https://www.theguardian.com/cities/2017/feb/27/sand-mining-global-environmental-crisis-never-heard

67 Beiser, V., 'Sand mining: the global environmental crisis you've probably never heard of', 27 February 2017, https://www.theguardian.com/cities/2017/feb/27/sand-mining-global-environmental-crisis-never-heard

68 http://www.bbc.com/capital/story/20160502-even-desert-city-dubai-imports-its-sand-this-is-why

69 Banyan, 'Such quantities of sand', 26 February 2015, *The Economist*, http://www.economist.com/news/asia/21645221-asias-mania-reclaiming-land-sea-spawns-mounting-problems-such-quantities-sand

70 http://www.huffingtonpost.com/entry/steve-bannon-apocalypse_us_5898f02ee4b040613138a951

71 http://www.huffingtonpost.com/entry/china-south-china-sea-sovereignty_us_588763f8e4b096b4a23483ab?

72 https://en.wikipedia.org/wiki/War_with_the_Newts

73 *Environmental Justice Atlas*, 'A planet in danger: The world

of Chevron', http://ejatlas.org/featured/chevronconflicts

74　Freedonia, 'World Industrial Silica Sand', http://www. freedoniagroup.com/industry-study/3237/world-industrial-silica-sand.htm

75　Fox., J., 'Gasland', documentary, https://en.wikipedia.org/ wiki/Gasland

76　Goldenberg, S. 'Greenpeace exposes sceptics hired to cast doubt on climate science', 8 December 2015, https://www. theguardian.com/environment/2015/dec/08/greenpeace-exposes-sceptics-cast-doubt-climate-science

77　Conway, E.M., Oreskes, N., *Merchants of Doubt. How a Handful of Scientists Obscured the Truth on Issues from Tobacco Smoke to Global Warming*, 2010, Bloomsbury Press

78　Brysse, K., Oreskes, N., O'Reilly, J., Oppenheimer, M., 'Climate change prediction: Erring on the side of least drama?', February 2013, Global Environmental Change, http://www.sciencedirect.com/science/article/pii/ S0959378012001215

79　Hansen, J., et al. 'Ice melt, sea level rise and superstorms: evidence from paleoclimate data, climate modeling, and modern observations that 2 °C global warming could be dangerous', 22 March 2016, http://www.atmos-chem-phys. net/16/3761/2016/

80　Branson, K., 'New finding shows climate change can happen in a geological instant', 7 October 2013, https://phys.org/ news/2013-10-climate-geological-instant.html

81　IPCC, 'Working Group II: Impacts, Adaptation and Vulnerability', 2001, http://www.ipcc.ch/ipccreports/tar/ wg2/index.php?idp=605

82　https://www.theguardian.com/environment/2016/aug/21/ arctic-will-be-ice-free-in-summer-next-year

83　*The Economist*, 'Energy firms and climate change. Unburnable fuels', 4 May 2013, http://www.economist. com/news/business/21577097-either-governments-are-not-serious-about-climate-change-or-fossil-fuel-firms-are

84   McKibben, B., 'A world at war', 15 August 2016, https://
      newrepublic.com/article/135684/declare-war-climate-
      change-mobilize-wwii. Oil Change International, 'The Sky's
      Limit', 22 September 2016, http://priceofoil.org/2016/09/22/
      the-skys-limit-report/

85   Howard, E., Parsons, J., 'Keep it in the ground climate
      campaign: the week in brief', 29 May 2015, http://www.
      theguardian.com/environment/keep-it-in-the-ground-
      blog/2015/may/29/keep-it-in-the-ground-climate-
      campaign-the-week-in-brief?CMP=ema-60

86   CIEL, '(Mis)calcuated risk and climate change. Are rating
      agencies repeating credit crisis mistakes?', May 2015, http://
      www.ciel.org/wp-content/uploads/2015/06/CIEL_CRA_
      Brief_24Jun2015.pdf#_blank

87   Oslo Principles on Global Climate Change Obligations.
      http://www.osloprinciples.org/

88   Steffen, A., 'Trump, Putin and the Pipelines to Nowhere',
      15 December 2016, https://medium.com/@AlexSteffen/
      trump-putin-and-the-pipelines-to-nowhere-742d745ce8fd#.
      izj3ffbie

89   Fairfin, 'De divestmentbeweging komt naar België', 12 May
      2015, http://www.fairfin.be/actueel/nieuws/2015/05/de-div
      estment-beweging-komt-naar-belgi%C3%AB

90   De klimaatzaak, http://klimaatzaak.eu/nl

91   Sassen, S., *Expulsions. Brutality and complexity in the global
      economy*, 2014, Harvard University Press

92   Wise, T., 'What Happened to the Biggest Land Grab in
      Africa?', Foodtank, https://foodtank.com/news/2014/12/
      what-happened-to-the-biggest-land-grab-in-africa-
      searching-for-prosavana-in/

93   Grainger, M., Geary, K., 'The New Forests Company
      and its Uganda Plantations', 22 September 2011, Oxfam,
      https://www.oxfam.org/sites/www.oxfam.org/files/file_

attachments/cs-new-forest-company-uganda-plantations-220911-en_4.pdf

94 Kron, J., 'In Scramble for Land, Group Says, Company Pushed Ugandans Out', 21 September 2011, *The New York Times*, http://www.nytimes.com/2011/09/22/world/africa/in-scramble-for-land-oxfam-says-ugandans-were-pushed-out.html?_r=1&

95 Vidal, J., 'Ugandan farmer: "My land gave me everything. Now I'm one of the poorest"', 22 September 2011, *The Guardian*, https://www.theguardian.com/environment/2011/sep/22/uganda-farmer-land-gave-me-everything McGroarty, P., 'Moves to Snap Up Land in Africa Draw Scrunity', 22 September 2011, *The Wall Street Journal*, http://www.wsj.com/articles/SB10001 424053111 90456390457658467341932875 Webb, M., 'Oxfam sounds Uganda land-grab warning', 22 September 2011, Aljazeera, http://www.aljazeera.com/video/africa/2011/09/201192 2111515150690.html

96 Lang, C., 'Ugandan farmers kicked off their land for New Forests Company's carbon project', 23 September 2011, *Redd Monitor*, http://www.redd-monitor.org/2011/09/23/ugandan-farmers-kicked-of-their-land-for-new-forests-companys-carbon-project/

97 FSC-Watch, 'How Accreditation Services International (FSC-ASI) allows certifiers to break FSC's rules and issue certificates to non-compliant companies', 16 March 2008, http://www.fsc-watch.org/archives/2008/03/16/How_Accreditation_Se

98 FSC-Watch, 'Former FSC boss admits core part of FSC system is a myth', 12 September 2014, https://fsc-watch.com/2014/09/16/former-fsc-boss-admits-core-part-of-fsc-system-is-a-myth/#more-762

99 Lang, C., 'Who watched the watchmen? RSPO's greenwashing and fraudulent reports exposed', 17

November 2015, *Redd Monitor*, http://www.redd-monitor. org/2015/11/17/who-watches-the-watchmen-rspos-green-washing-and-fraudulent-reports/

100 Broers, L., Lecluyse, A., 'Duurzaam op papier', 25 August 2010, http:// www.mo.be/magazine/september-2010/duurza am-op-papier en http://www.duurzaamoppapier.be/

101 Lang, C., 'Modern slavery found in RSPO member Felda Global Ventures' oil palm plantations', 5 August 2015, http://www.redd-monitor.org/2015/08/05/modern-slavery-found-in-rspo-member-felda-global-ventures-oil-palm-plantations/

102 Broers, L., 15 June 2012, 'WWF onder vuur: "De panda duldt geen kritiek!"', http://www.mo.be/opinie/wwf-onder-vuur-de-panda-duldt-geen-kritiek

103 Transport & Environment, 'Biodiesel increasing EU transport emissions by 4% instead of cutting CO_2', 4 May 2016, https://www.transportenvironment.org/news/biodiesel-increasing-eu-transport-emissions-4-instead-cutting-co2

104 EJOLT, 'International Day of Forests: Defining Forests by their true meaning!', 21 March 2014, http://www.ejolt. org/2014/03/international-day-of-forests-defining-forests-by-their-true-meaning/

105 FERN, 'EU consumption and illegal deforestation', 16 March 2015, http://www.fern.org/illegalconsumption

106 FERN, 'Stolen Goods: The EU's complicity in illegal tropical deforestation', 17 March 2015, http://www.fern. org/stolengoods

107 Pearce, F., *When the rivers run dry. Water, the defining crisis of the twenty-first century*, 2006, Beacon Press.

108 Europa Press, 'Fiscalía pide imputar al delegado de Medio Ambiente en Almería', 21 April 2014, http://www. europapress.es/andalucia/almeria-00350/noticia-fiscal-pide-imputar-delegado-medio-ambiente-cambio-uso-

suelo-forestal-agricola-20150420184825.html

109 *Canal Sur News*, 'The spring of Los Molinos del Rio Aguas is going dry!', 18 November 2014, https://www.youtube.com/watch?v=7YFJ3L1BDBM

110 'Boletín Official de la Junta de Andalucía', 23 December 2016, http:// www.juntadeandalucia.es/eboja/2016/245/BOJA16-245-00099-21925- 03_00104550.pdf

111 Singh, S., 'Local Governance and Environment Investments in Hiware Bazar, India', http://www.ceecec.net/case-studies/local-governance-and-environment-investments-in-hiware-bazar/

112 Martinez Alier, J., Temper, L., 'Is India Too Poor to be Green?', 28 April– 4 May 2007, vol.42, nr. 17, *Economic and Political Weekly*, https://www. academia.edu/25851029/Is_India_Too_Poor_To_Be_Green

113 *Pamir Times*, 'Glaciers Shape Lives in Upper Hunza', 3 June 2015, http:// pamirtimes.net/2015/06/03/glaciers-shape-lives-in-upper-hunza/

114 K.M. Lau, W., Kim, K., 'The 2010 Pakistan Flood and Russian Heat Wave: Teleconnection of Hydrometeorological Extremes', 1 February 2012, *Journal of Hydrometeorology*, http:// journals.ametsoc.org/doi/pdf/10.1175/JHM-D-11-016.1

115 Hornborg, A., *Global Magic: Technologies of Appropriation from Ancient Rome to Wall Street*, 2016, Springer

116 Schaffartzik, A., Mayer, A., Gingrich, S., Eisenmenger, N., Loy, C., Krausmann, F., 'The global metabolic transition: Regional patterns and trends of global material flows', 1950–2010', May 2014, *Glob Environ Change*, https://www. ncbi.nlm.nih.gov/pmc/articles/PMC4375797/

117 Mayer, A., Haas, W., 'Cumulative material flows provide indicators to quantify the ecological debt', 2016, *Journal of Political Ecology*, 23: 350-363, http://jpe.library.arizona.edu/volume_23/MayerandHaas.pdf

118 Overbeek, W., 'An overview of industrial tree plantations in

the global South. Recommendations for policy makers', 12 September 2012, EJOLT, http:// www.ejolt.org/wordpress/ wp-content/uploads/2012/09/001_Tree-plantations.pdf

119 Mayer, A., Schaffartzik, A., Haas, W., Rojas Sepúlveda, A., 'Patterns of global biomass trade. Implications for food sovereignty and socio-environmental conflicts', 2015, EJOLT Report nr.20, http://www.ejolt.org/2015/03/global-biomass-robbery/

120 De Schutter, L., Lutter, S., 'The true cost of consumption: the EU's land footprint', July 2016, http://www.foeeurope. org/true-cost-consumption-land-footprint-report

121 Krausmann, F., et al., 'Global human appropriation of net primary production doubled in the 20th century', 5 May 2013, Proceedings of the National Academy of Science, http://www.pnas.org/content/110/25/10324.full.pdf

122 *Het Nieuwsblad*, 'Er is onvoldoende landbouwgrond om ons lokaal te voeden', 11 February 2014, http://www. nieuwsblad.be/cnt/dmf20140211_00974566

123 European Environmental Bureau, 'Advancing resource efficiency in Europe', March 2014, http://www. eeb.org/EEB/?LinkServID=4E9BB68D-5056-B741-DBCCE36ABD15F02F

124 Bond Beter Leefmilieu, 'Je "lokaal stukje vlees" zit vol importsoja', 3 March 2017, https://www. bondbeterleefmilieu.be/artikel/je-lokaal-stukje-vlees- zit-vol-importsoja? utm_source=MailingList&utm_medium =email&utm_campaign=Beleidsbabbel+170302

125 Milman., O., 'Rate of environmental degradation puts life on Earth at risk, say scientists', 15 January 2015, https:// www.theguardian.com/environ ment/2015/jan/15/rate-of-environmental-degradation-puts-life-on-earth-at-risk-say-scientists?CMP=share_btn_fb

126 Steffen, W., et al., 'Planetary boundaries: Guiding human development on a changing planet', 13 February 2015, *Science*,

http://science.sciencemag.org/ content/347/6223/1259855

127 Monbiot, G., 'We're treating soil like dirt. It's a fatal mistake, as our lives depend on it', 25 March 2015, https://www. theguardian.com/commentisfree/2015/mar/25/treating-soil-like-dirt-fatal-mistake-human-life?CMP=share_btn_fb

128 Foucher, A., et al., 'Increase in soil erosion after agricultural intensification: Evidence from a lowland basin in France', September 2014, *Anthropocene*, http://www.sciencedirect. com/science/article/pii/S221330541500003X

129 UNCTAD, 'Trade and Environment review 2013. Wake up before it's too late', 18 September 2013, http://unctad.org/ en/publicationslibrary/ditc-ted2012d3_en.pdf

130 Edmonson, J.L, et al., 'Urban cultivation in allotments maintains soil qualities adversely affected by conventional agriculture', 24 April 2014, *Journal of Applied Ecology*, http:// onlinelibrary.wiley.com/doi/10.1111/13652664.12254/full

131 *National Geographic*, 'Our dwindling food variety', http:// ngm.nationalgeographic.com/2011/07/food-ark/food-variety-graphic

132 NO REDD in Africa network, 'The Worst REDD-type projects (i) in Africa: Continent Grab for Carbon Colonialism', http:// no-redd-africa.org/index.php/16-redd-   players/84-the-worst-redd-type-projects-in-africa-continent-grab-for-carbon-colonialism#_edn35

133 Bond, P., Sharife, K., Allen, F., Amisi, B., Brunner, K, Castel-Branco, R., Dorsey, D., Gambirazzio, G., Hathaway, T., Nel, A., Nham, W., 'The CDM cannot deliver the money to Africa. Why the Clean Development Mechanism won't save the planet from climate change, and how African civil society is resisting', 2012, EJOLT Report, nr.2, 120p. http://www. ejolt.org/2012/12/the-cdm-cannot-deliver-the-money-to-africa-why-the-carbon-trading-gamble-won%E2%80%99t-save-the-planet-from-climate-change-and-how-african-civil-society-is-resisting/

134 *Het Laatste Nieuws*, 'Vlaanderen koopt 20 miljoen ton schone lucht', 8 January 2014, http://www.hln.be/hln/nl/2764/milieu/article/detail/1769960/2014/   01/08/Vlaanderen-koopt-20-miljoen-ton-schone-lucht.dhtml

135 *De Morgen*, 'De prijs van "schone" lucht is veel te hoog. Drie onderzoekers van de KU Leuven vinden handel in "schone" lucht een maat voor niets', 28 November 2011, http://www.demorgen.be/plus/de-prijs-van-schone-lucht-is-veel-te-hoog-b-1412188626667/

136 Meynen, N., 'Vlaamse regering investeert massaal in klimatologisch piramidespel', 8 January 2014, http://www.mo.be/opinie/vlaamse-regering-investeert-massaal-klimatologisch-piramidespel

137 FERN, 'Scrap the ETS Coalition', 23 April 2014, http://www.fern.org/book/trading-carbon/scrap-ets-coalition

138 Kenis, A., Lievens, M., *De mythe van de groene economie*, 2012, EPO.

139 Meynen, N., 'Environmental Justice and Ecological Debt in Belgium: The UMICORE case', 2012, in: *Ecological Economics from the Ground Up*, Routledge

140 Mertens,V., 'Baanbrekend onderzoek brengt ecologische schuld in kaart',VMX, http://www.vmx.be/baanbrekend-onderzoek-brengt-ecologische-schuld-kaart

141 Stassijns, J., 'Hannah (2) heeft gevaarlijk gehalte lood in bloed na werken Umicore', 27 January 2017, *Gazet van Antwerpen*, http://www.gva.be/cnt/dmf20170126_02698042/hannah-2-heeft-gevaarlijk-gehalte-lood-in-bloed-na-werken-umicore

142 Roberts, D., 'None of the world's top industries would be pro table if they paid for the natural capital they use', 17 April 2013, http://grist.org/business-technology/none-of-the-worlds-top-industries-would-be-profitable-if-they-paid-for-the-natural-capital-they-use/

143 Do Something, '11 Facts about E-Waste', https://www.

dosomething.org/facts/11-facts-about-e-waste

144 ARTE, La tragédie électronique (documentaire), 30 September 2015, http://future.arte.tv/fr/la-tragedie-electronique.

145 Electronics TakeBack Coalition, 'Facts and Figures on E-Waste and Recycling', http://www.electronicstakeback.com/wp-content/uploads/Facts_ and_Figures_on_EWaste_and_Recycling.pdf

146 Jardim, E., 'What 10 years of smartphone use means for the planet', 27 February 2017, http://www.greenpeace.org/international/en/news/Blogs/makingwaves/smarphones-planet-toxicwaste/blog/58828/

147 Slade, G., 'iWaste. Ten thousand songs in your pocket. Ten thousand years in a land ll.', March / April 2007, http://www.motherjones.com/environment/2007/03/iwaste

148 Vlaamse Overheid, '22,4 kg', http://do.vlaanderen.be/224-kg

149 UNEP, 'Sustainable Innovation and Technology Transfer Industrial Sector Studies. Recycling – From E-Waste to Resources', 2009, http://www.unep.org/pdf/Recycling_From_e-waste_to_resources.pdf

150 Reports from Earth, 'Designed to fail: planned obsolescence in printers – tricks to fix them!', 5 August 2011, www.reportsfromearth.com/155/designed-to-fail-planned-obsolescence-in-printers-tricks-to-fix-them

151 https://www.ifixit.com/ and https://www.repairclinic.com/

152 The Dirty Little Secret Of Inkjet Printers, 10 November 2008, https://www.youtube.com/watch?v=ycD4XkUtbIw

153 https://repaircafe.org

154 http://produitspourlavie.org/

155 http://www.murks-nein-danke.de/blog/

156 Van Eeckhout, L., 'L'obsolescence programmée des produits désormais sanctionnée', 15 October 2014, http://www.lemonde.fr/planete/article/2014/10/15/l-

obsolescence-programmee-des-produits-desormais-sanc-tionnee_4506580_3244.html

157 Haas, W., Krausmann, F. Wiedenhofer, D. and Heinz, M., 'How Circular is the Global Economy? A Sociometabolic Analysis', in: *Social Ecology: Society-nature Relations across Time and Space* (red.) Haberl, H., Fischer-Kowalski, M., Krausmann, F. and Winiwarter V., 2016, Springer.

158 Oxford University Press USA, 'Current climate change models understate the problem, scientists argue', 8 February 2017, https://www.sciencedaily. com/releases/ 2017/02/170208111626.htm

159 Schaffartzik, A., Mayer, A., Gingrich, S., Eisenmenger, N., Loy, C., Krausmann, F., 'The global metabolic transition: Regional patterns and trends of global material flows, 1950–2010', May 2014, Glob Environ Change, https://www. ncbi.nlm.nih.gov/pmc/articles/PMC4375797/

160 Legambiente, 'Ecomafia 2009', 2010, http://www.legam biente.it/sites/default/ les/docs/Ecomafia_2009_0000001861. pdf

161 Legambiente, 'Legambiente presenta Ecomafia 2016, le storie e i numeri della criminalità ambientale in Italia', 5 July 2016, http://www.legambiente.it/contenuti/comunicati/ legambiente-presenta-ecomafia-2016-le-storie-e-i-numeri-della-criminalita-ambie

162 https://www.legambiente.it/contenuti/comunicati/ legambiente-presenta-ecomafia-2016-le-storie-e-i-numeri-della-criminalita-ambie

163 Koomen, W., De Gifcirkel (Documentary on the waste crisis in Campania), 2015

164 UN OHCHR, 'Ten years on, the survivors of illegal toxic waste dumping in Côte d'Ivoire remain in the dark', 19 August 2016, http://www.ohchr.org/EN/NewsEvents/Pages /DisplayNews.aspx?NewsID=20384&LangID=E

165 Public Eye, 'Dirty Diesel. How Swiss traders Flood Africa

with Toxic Fuels', September 2016, https://www.publiceye.
ch/fileadmin/files/documents/2016_DirtyDiesel_A-Public-
Eye-Investigation_final.pdf

166 The Basel Convention on the Control of Transboundary
Movements of Hazardous Wastes and their Disposal, the
Rio Declaration on Environment and Development, the ILO
138 on the Minimum Age Convention, the ILO 182 on the
Worst Forms of Child Labor, the Universal Declaration of
Human Rights, the International Covenant on Economic,
Social and Cultural Rights and the OECD Guidelines for
Multinational Enterprises.

167 Tange, N., 'Belgische rederij liet schip slopen op omstreden
werf in Bangladesh', 6 May 2016, http://www.standaard.be/
cnt/dmf20160505_02275326

168 *De Standaard*, 'Muilezels van het water winnen terrein',
10 February, http://www.standaard.be/cnt/dmf20170209
_02722902

169 WTO, 'World Trade Report 2013. Factors shaping the future
of world trade', 2013, https://www.wto.org/english/res_e/
booksp_e/wtr13-2b_e.pdf

170 Meynen, N., 'New scientific insights on ecologically unequal
trade', 23 November 2016, http://www.theecologist.
org/blogs_and_comments/commentators/2988366/new_
scientific_insights_on_ecologically_unequal_trade.html

171 Pearce, F., 'How 16 ships create as much pollution as all
the cars in the world', 21 November 2009, http://www.
dailymail.co.uk/sciencetech/article-1229857/How-16-ships-
create-pollution-cars-world.html

172 In 1970 total value of exports was $317000 million,
in 2016 it was $15985000 million http://stat.wto.org/
StatisticalProgram/WSDBViewData.aspx?Language=E

173 Mertens., J., 'Laten we het woord "bescherming" heroveren',
6 February 2017, http://www.mo.be/column/laten-we-het-
woord-bescherming-heroveren

174 CEO, 'The great CETA swindle', November 2016, https://
corporateeurope.org/sites/default/files/attachments/great-
ceta-swindle.pdf

175 Stop TTIP, 'Legal Statement on investment protection in
TTIP and CETA', 17 October 2016, https://stop-ttip.org/
blog/legal-statement-on-investment-protection-in-ttip-
and-ceta/

176 Hamby, C., 'The court that rules the world', 28 August 2016,
https://www.buzzfeed.com/chrishamby/super-court?utm_
term=.rdE2QGrO6W#.yqgQeL2yvr

177 De Grauwe, P., 'Om deze redenen vind ik dat
vrijhandelsakkoorden in de ijskast moeten worden gestopt',
24 oktober 2016, http://www.demorgen.be/opinie/om-deze-
redenen-vind-ik-dat-vrijhandelsakkoorden-in-de-ijskast-
moeten-worden-gestopt-be7666c3/

178 Holslag, J., 'Wat er écht mis is met het handelsverdrag met
Canada', 19 October 2016, http://www.jonathanholslag.be/
blog/wat-er-echt-verkeerd-is-aan-het-handelsverdrag-met-
canada/

179 Harmsen, V., 'TTIP en CETA: de grote verdwijntruc van
Europese milieustandaarden', 2016, http://www.mo.be/
analyse/ttip-en-ceta-de-grote-verdwijntruc-van-europese-
milieustandaarden

180 Horel, S., 'A toxic Affair', May 2015, CEO, https://
corporateeurope.org/sites/default/files/toxic_lobby_edc.
pdf

181 CEO, 'Leaked document shows EU is going for a trade
deal that will weaken financial regulation', 1 July 2014,
https://corporateeurope.org/financial-lobby/2014/07/
leaked-document-shows-eu-going-trade-deal-will-weaken-
financial-regulation#footnoteref15_hjy1umk

182 EEA, 'Air Pollution', http://www.eea.europa.eu/themes/air

183 Van de Weijer, B., 'Niets geleerd uit dieselgate: gat tussen
uitstootbelofte en werkelijk verbruik steeds groter', 21

December 2016, http://www.demor- gen.be/buitenland/-niets-geleerd-uit-dieselgate-gat-tussen-uitstootbelofte-en-werkelijk-verbruik-steeds-groter-bac681c0/

184 Belga, 'Europa had Dieselgate eerder kunnen aanpakken', 28 February 2017, http://www.standaard.be/cnt/dmf201 70228_02755604

185 Higgins, A., 'Volkswagen Scandal Highlights European Stalling on New Emissions Tests', 28 September 2015, http://www.nytimes.com/2015/09/29/world/europe/volkswagen-scandal-highlights-european-stalling-on-new-emissions-tests.html?_r=2

186 Oxfam, 'Tax Battles. The dangerous global Race to the Bottom on Corporate Tax', 12 December 2016, http://www.oxfamnovib.nl/Redactie/Downloads/Rapporten/20161212%20bp-race-to-bottom-corporate-tax-121216-en-EMBARGO.pdf

187 Bowers, S., 'Jean-Claude Juncker blocked EU curbs on tax avoidance, cables show', 1 January 2017, https://www.theguardian.com/business/2017/jan/01/jean-claude-juncker-blocked-eu-curbs-on-tax-avoidance-cables-show

188 Verhaeghe, P., 'Neoliberalism has brought out the worst in us', 29 September 2014, https://www.theguardian.com/commentisfree/2014/sep/29/neoliberalism-economic-system-ethics-personality-psychopathicsthic

189 Free Exchange, 'Are economists erring on climate change?', 9 February 2011, http://www.economist.com/blogs/freee xchange/2011/02/climate_policy

190 Wikipedia, 'The Limits to Growth', https://en.wikipedia.org/wiki/The_ Limits_to_Growth

191 EEA, 'Late lessons from early warnings: science, precaution, innovation', 22 January 2013, http://www.eea.europa.eu/publications/late-lessons-2

192 Monbiot, G., 'Cop-Out', 15 December 2015, http://www.monbiot.com/2015/ 12/15/cop-out/

193 Milman, O., 'James Hansen, father of climate change awareness, calls Paris talks "a fraud"', 12 December 2015, https://www.theguardian.com/environment/2015/dec/12/james-hansen-climate-change-paris-talks-fraud

194 Vandaele, J., 'Regering start debat over CO_2-taks maar of er ook daden volgen, is onzeker', 24 January 2017, http://www.mo.be/nieuws/regering-start-debat-over-co2-taks-maar-er-ook-daden-volgen-onzeker

195 Rombaut, M., 'De Flat Earth Society groeit: waarom geloven steeds meer mensen dat de aarde plat is?', 6 March 2017, http://m.newsmonkey.be/article/76615

196 https://www.thelocal.fr/20180108/from-flat-earth-theory-to-the-moon-landings-what-the-french-think-of-conspiracy-theories

197 Steffen, W., et al., 'The trajectory of the Anthropocene: The Great Acceleration', 2015, *The Anthropocene Review*, http://anr.sagepub.com/content/2/1/81.full.pdf+html

198 MIT, 'Study: Technological progress alone won't stem resource use', 19 January 2017, http://news.mit.edu/2017/technological-progress-alone-stem-consumption-materials-0119#.WIi8Z2rXukE.facebook

199 Kallis, G., In defense of degrowth. Opinions and minifestos, 2017 (e-book)

200 Monbiot, G., 'Growth: the destructive god that can never be appeased', 18 November 2014, https://www.theguardian.com/commentisfree/2014/nov/18/growth-destructive-economic-expansion-financial-crisis.

201 Joossen, A., 'Club van Rome krijgt gelijk: "Wereld vergaat nog deze eeuw"', 3 September 2014, http://www.demorgen.be/wetenschap/club-van-rome-krijgt-gelijk-wereld-vergaat-nog-deze-eeuw-b3151c3f/

202 Ahmed, N., 'Scientists vindicate "Limits to Growth" – urge investment in "circular economy"', 4 June 2014, https://www.theguardian.com/environment/earth-insight/2014/

jun/04/scientists-limits-to-growth-vindicated-investment-transition-circular-economy

203 Pope Francis, 'Laudato Si', 18 June 2015, http://w2.vatican. va/content/francesco/en/encyclicals/documents/papa-francesco_20150524_enciclica-laudato-si.html

204 https://theecologist.org/2018/mar/27/rise-and-future-degrowth-movement

205 Martinez Alier, J., Temper, L., Del Bene, D., Scheidel, A., 'Is there a global environmental justice movement?', 2016, *The Journal of Peasant Studies*, http://www.tandfonline.com/doi/full/10.1080/03066150.2016.1141198

206 Klein, N., *This Changes Everything: Capitalism vs the Climate*, 2014, Simon & Schuster

207 Wikipedia, 'Genuine progress indicator', https://en.wikipedia.org/wiki/Genuine_progress_indicator

208 Kubiszewski, I., et al, 'Beyond GDP: Measuring and achieving global genuine progress', 2013, *Ecological Economics*, http://www.ajfand.net/Volume13/No5/Reprint-Beyond%20GDP.pdf%20

209 Sassen, S., *Expulsions. Brutality and complexity in the global economy*, 2014, Harvard University Press

210 Sassen, S., idem

211 Helfrich, S., Bollier, D., *The Wealth of the Commons. A world beyond market and state*, 2012, The Levellers Press.

212 Demailly, D., Chancel, L., Waisman, H., Guivarch, C., 'A post-growth society for the 21st century. Does prosperity have to wait for the return of economic growth?', 13 November 2013, http://www.iddri.org/Publications/Collections/Analyses/Study0813_DD%20et%20al._post-growth%20society.pdf.

213 OXFAM, 'An Economy for the 99%', January 2017, https://www.oxfam.org/sites/www.oxfam.org/files/ file_attachments/bp-economy-for-99-percent-160117-en.pdf

214 Piketty, T., *Capital in the twenty-first century*, 2014, Harvard

University Press.

215 Decreus, T., Callewaert, C., *Dit is morgen*, 2015, EPO.

216 NEF, *21 hours. Why a shorter working week can help us all to flourish in the 21st century*, 2010.

217 UN, 'The Future we want: final document of the Rio+20 Conference', 25 June 2012, http://rio20.net/en/iniciativas/the-future-we-want-final-document-of-the-rio20-conference/

218 Falkenberg, K., 'Sustainability Now! A European Vision for Sustainability', 20 July 2016, https://ec.europa.eu/epsc/sites/epsc/files/strategic_note_issue_18.pdf.

219 *National Geographic*, 'State of the Earth', 22 December 2009.

220 Myllyvirta, L., 'Mapped: The coal power plants China just suspended', 18 January 2017, http://energydesk.greenpeace.org/2017/01/18/china-climate-leader-coal-davos-xi-jinping/

221 *The Economist*, 'Transparency in the haze', 8 February 2014, http://www.economist.com/news/china/21595927-government-takes-steps-towards-more-openness-transparency-haze

222 *The Economist*, 'Right to know', 3 May 2014, http://www.economist.com/news/china/21601564-leaders-discover-some-transparency-can-help-make-society-more-stable-right-know

223 *The Economist*, 'Green Teeth', 17 May 2014, http://www.economist.com/news/china/21602286-government-amends-its-environmental-law-green-teeth

224 *The Economist*, 'Saving fish and baring teeth', 18 April 2015, http://www.economist.com/news/china/21648687-new-environment-minister-displays-his-appetite-taking-polluters-saving-fish-and-baring

225 *The Economist*, 'A small breath of fresh air', 8 February 2014, http://www.economist.com/news/leaders/21595903-government-gives-its-davids-sling-use-against-polluting-goliaths-small-breath-fresh

226　Jing, C., *Under the Dome* (documentary), https://en.wikipedia. org/wiki/Under_the_Dome_(film)

227　Meynen, N., *Nepal. Nieuwe wegen in de Himalaya*, 2016, EPO

228　Kingdom of Bhutan, 'Happiness: Towards a New Development Paradigm', 2013, http://www.new developmentparadigm.bt/wp-content/uploads/2014/10/ HappinessTowardsANewDevelopmentParadigm.pdf

229　Hayden, A., 'Bhutan: Blazing a Trail to a Postgrowth Future? Or Stepping on the Treadmill of Production?', 14 April 2015, *The Journal of Environment & Development*, http://jed. sagepub.com/content/early/2015/04/14/1070496515579199

230　Mounk, Y., Foa, R.S., 'The Signs of Deconsolidation', January 2017, *Journal of Democracy*, http://www.journalofdemocracy. org/article/signs-deconsolidation

231　Taub, A., 'How Stable Are Democracies? "Warning Signs Are Flashing Red"', 29 November 2016, http://www. nytimes.com/2016/11/29/world/americas/western-liberal- democracy.html?smid=tw-nytimes&smtyp=-　　　　cur&_ r=2&referer=https%3A%2F%2Ft.co%2FjobRTvFrPO

232　De Ceulaer, J., 'De waarheid bestaat wél (maar ze doet er niet toe). Waarom feiten niet zo belangrijk zijn in de politiek', 3 March 2017, http://www.demorgen.be/politiek/de- waarheid-bestaat-wel-maar-ze-doet-er-niet-toe-b853ab9d/

233　Watts, J., "We have 12 years to limit climate change catastrophe, warns UN", 8 October 2018, https://www. theguardian.com/environment/2018/oct/08/global- warming-must-not-exceed-15c-warns-landmark-un-report

234　Johnston, I., 'Record number of oil and gas firms go bust as renewable energy revolution begins to bite', 3 January 2017, http://www.independent.co.uk/environment/oil-gas- firms-industry-bust-renewable-energy-revolution-biofuel- solar-panel-wind-power-opec-saudi-a7507016.html

Further reading

The endnotes in this book refer to the sources I used directly, but neither does that list cover the full picture of my sources, not does it give the reader a good overview of what to read next. Here I'm listing books and longer texts that have inspired me in the broader sense. While this list is also far from complete, it can be interpreted as a recommended reading list for those who wish to dig deeper after reading this book. In alphabetical order:

Diamond, J., *Collapse. How societies choose to fail or succeed*, 2006.

Holemans, D., Freedom & Security. Towards a social-ecological society (in Dutch), 2016. English booklet version available at www.oikos.be

Jackson, T., *Prosperity without growth: Economics for a Finite Planet*, 2009.

Kenis, A., Lievens, M., editor(s), *The Myth of the Green Economy*, 2015.

Klein, N., *This Changes Everything: Capitalism vs the Climate*, 2014.

Klein, N., *NO is not Enough,* 2017.

Pope Francis, *'Laudato Si'*, 2015.

Pearce, F., *When the rivers run dry. Water, the defining crisis of the twenty-first century*, 2006.

Sassen, S., *Expulsions. Brutality and complexity in the global economy*, 2014.

Shrivastava, A., Kothari, A., *Churning the Earth. The making of global India*, 2012.

Stiglitz, J., *Making Globalization Work*, 2006.

Wilkinson, R., Pickett, K., *The spirit level. Why more equal societies almost always do better*, 2009.

CULTURE, SOCIETY & POLITICS

Contemporary culture has eliminated the concept and public figure of the intellectual. A cretinous anti-intellectualism presides, cheer-led by hacks in the pay of multinational corporations who reassure their bored readers that there is no need to rouse themselves from their stupor. Zer0 Books knows that another kind of discourse – intellectual without being academic, popular without being populist – is not only possible: it is already flourishing. Zer0 is convinced that in the unthinking, blandly consensual culture in which we live, critical and engaged theoretical reflection is more important than ever before.

If you have enjoyed this book, why not tell other readers by posting a review on your preferred book site.

Recent bestsellers from Zero Books are:

In the Dust of This Planet
Horror of Philosophy vol. 1
Eugene Thacker
In the first of a series of three books on the Horror of Philosophy,
In the Dust of This Planet offers the genre of horror as a way of
thinking about the unthinkable.
Paperback: 978-1-84694-676-9 ebook: 978-1-78099-010-1

Capitalist Realism
Is there no alternative?
Mark Fisher
An analysis of the ways in which capitalism has presented itself
as the only realistic political-economic system.
Paperback: 978-1-84694-317-1 ebook: 978-1-78099-734-6

Rebel Rebel
Chris O'Leary
David Bowie: every single song. Everything you want to know,
everything you didn't know.
Paperback: 978-1-78099-244-0 ebook: 978-1-78099-713-1

Cartographies of the Absolute
Alberto Toscano, Jeff Kinkle
An aesthetics of the economy for the twenty-first century.
Paperback: 978-1-78099-275-4 ebook: 978-1-78279-973-3

Poor but Sexy
Culture Clashes in Europe East and West
Agata Pyzik
How the East stayed East and the West stayed West.
Paperback: 978-1-78099-394-2 ebook: 978-1-78099-395-9

Malign Velocities
Accelerationism and Capitalism
Benjamin Noys
Long listed for the Bread and Roses Prize 2015, *Malign Velocities* argues against the need for speed, tracking acceleration as the symptom of the ongoing crises of capitalism.
Paperback: 978-1-78279-300-7 ebook: 978-1-78279-299-4

Meat Market
Female Flesh under Capitalism
Laurie Penny
A feminist dissection of women's bodies as the fleshy fulcrum of capitalist cannibalism, whereby women are both consumers and consumed.
Paperback: 978-1-84694-521-2 ebook: 978-1-84694-782-7

Romeo and Juliet in Palestine
Teaching Under Occupation
Tom Sperlinger
Life in the West Bank, the nature of pedagogy and the role of a university under occupation.
Paperback: 978-1-78279-637-4 ebook: 978-1-78279-636-7

Sweetening the Pill
or How We Got Hooked on Hormonal Birth Control
Holly Grigg-Spall
Has contraception liberated or oppressed women? *Sweetening the Pill* breaks the silence on the dark side of hormonal contraception.
Paperback: 978-1-78099-607-3 ebook: 978-1-78099-608-0

Why Are We The Good Guys?
Reclaiming your Mind from the Delusions of Propaganda
David Cromwell
A provocative challenge to the standard ideology that Western
power is a benevolent force in the world.
Paperback: 978-1-78099-365-2 ebook: 978-1-78099-366-9

Readers of ebooks can buy or view any of these bestsellers by
clicking on the live link in the title. Most titles are published
in paperback and as an ebook. Paperbacks are available in
traditional bookshops. Both print and ebook formats are available
online.
Find more titles and sign up to our readers' newsletter
at http://www.johnhuntpublishing.com/culture-and-politics
Follow us on Facebook
at https://www.facebook.com/ZeroBooks
and Twitter at https://twitter.com/Zer0Books